AN INTRODUCTION TO
TENSOR CALCULUS AND RELATIVITY

An Introduction to Tensor Calculus and Relativity

(METHUEN'S MONOGRAPHS
ON PHYSICAL SUBJECTS)

Derek F. Lawden,

SC.D., F.R.S.N.Z.

*Professor of Mathematics
University of Canterbury, N.Z.*

METHUEN & CO LTD
and
SCIENCE PAPERBACKS

11 NEW FETTER LANE, LONDON E.C.4

First published by Methuen & Co Ltd 1962
Second edition 1967
Reprinted 1968
S.B.N. 416 43140 2
2.2

First published as a Science Paperback 1967
Reprinted 1968
S.B.N. 412 20370 7

© 1967 *by Derek F. Lawden*
Printed in Great Britain by
Fletcher & Son Ltd, Norwich

This book is available in both hardbound and paperback editions. The paperback edition is sold subject to the condition that it shall not, by way of trade or otherwise, be lent, resold, hired out, or otherwise circulated without the publisher's prior consent in any form of binding or cover other than that in which it is published and without a similar condition including this condition being imposed on the subsequent purchaser.

Distributed in the U.S.A.
by Barnes & Noble Inc.

Contents

Preface	*page* ix
1. Special Principle of Relativity. Lorentz Transformations	1
1. Newton's laws of motion	1
2. Covariance of the laws of motion	4
3. Special principle of relativity	5
4. Lorentz transformations. Minkowski space-time	7
5. The special Lorentz transformation	11
6. Fitzgerald contraction. Time dilation	14
7. Spacelike and timelike intervals. Light cone	17
Exercises 1	20
2. Orthogonal Transformations. Cartesian Tensors	23
8. Orthogonal transformations	23
9. Repeated index summation convention	25
10. Rectangular Cartesian tensors	26
11. Invariants. Gradients. Derivatives of tensors	29
12. Contraction. Scalar product. Divergence	30
13. Tensor densities	32
14. Vector products. Curl	33
Exercises 2	35
3. Special Relativity Mechanics	38
15. The velocity vector	38
16. Mass and momentum	41
17. The force vector. Energy	43
18. Lorentz transformation equations for force	47
19. Motion with variable proper mass	48
20. Lagrange's and Hamilton's equations	50
Exercises 3	51

4. Special Relativity Electrodynamics *page* 59

21. 4-Current density 59
22. 4-Vector potential 60
23. The field tensor 62
24. Lorentz transformations of electric and magnetic intensities 65
25. The Lorentz force 67
26. Force density 68
27. The energy–momentum tensor for an electromagnetic field 69
28. Equations of motion of a charge flow 74
    Exercises 4 77

5. General Tensor Calculus. Riemannian Space 81

29. Generalized N-dimensional spaces 81
30. Contravariant and covariant tensors 85
31. The quotient theorem. Conjugate tensors 92
32. Relative tensors and tensor densities 94
33. Covariant derivatives. Parallel displacement. Affine connection 96
34. Transformation of an affinity 99
35. Covariant derivatives of tensors 101
36. Covariant differentiation of relative tensors 104
37. The Riemann-Christoffel curvature tensor 108
38. Geodesic coordinates. The Bianchi identities 113
39. Metrical connection. Raising and lowering of indices 115
40. Scalar products. Magnitudes of vectors 117
41. The Christoffel symbols. Metric affinity 118
42. The covariant curvature tensor 121
43. Divergence. The Laplacian. Einstein's tensor 123
44. Geodesics 125
    Exercises 5 129

6. General Theory of Relativity 137

45. Principle of equivalence 137
46. Metric in a gravitational field 141

47. Motion of a free particle in a gravitational field *page*	145
48. Einstein's law of gravitation	147
49. Acceleration of a particle in a weak gravitational field	150
50. Newton's law of gravitation	153
51. Metrics with spherical symmetry	155
52. Schwarzschild's solution	158
53. Planetary orbits	160
54. Gravitational deflection of a light ray	166
55. Gravitational displacement of spectral lines	168
56. Maxwell's equations in a gravitational field	171
Exercises 6	173
Appendix Bibliography	182
Index	183

Preface

Now that half a century has elapsed since the special and general theories of relativity were constructed, it is possible to perceive more clearly their true significance for the development of mathematical physics as a whole. Although these theories possessed a decidedly revolutionary appearance when they were announced, it has now become clear that they represented the natural termination for the classical theories of mechanics and electromagnetism, rather than a break with these systems of ideas and the inception of a new line of thought. In this respect, the relativity theories are to be contrasted with that other great achievement of modern theoretical physics, namely quantum mechanics, based as this is upon principles which are completely at variance with those which were fundamental for Newtonian mechanics. That relativity theory has proved to be somewhat sterile by comparison with the enormous fertility of the new ideas introduced in quantum theory may, perhaps, be partly accounted for by this difference. However, the circumstance that the applications of relativity theory are chiefly to problems in the fields of astronomy and cosmology, fields which until recently have not received great attention from physicists due to the difficulty of experimentation, whereas quantum mechanics is associated with phenomena of atomic dimensions more easily investigated in the laboratory, may also have a bearing upon the matter. Now that interplanetary and even interstellar instrumented probes are approaching realization, it may be that interest in cosmological problems will be stimulated and that significant advances from the position reached by relativity theory some decades ago will result.

However, relativity theory's present status is as the culmination of the classical theories of mechanics and electromagnetism and, since much time is normally spent during undergraduate courses giving a detailed account of these classical theories, it has always seemed to me improper that such courses should be terminated without a description being given of the very natural and illuminating form into which it is possible to cast these theories by use of the device, basic for relativity theory, of mapping events on a four-dimensional space-time manifold. Such an introductory course of lectures on relativity theory helps to clarify the principles upon which the classical theories are based and is far more rewarding from the student's point of view than is a course devoted to the solution of intricate problems of statics and dynamics. By some sacrifice of the time usually devoted to conventional techniques of classical mechanics and electromagnetism, it is possible to provide an introductory course of relativity theory suitable for students in their third honours year at the university who, judging from our experience at Canterbury, are at this stage sufficiently mature mathematically to appreciate the great beauty of this system of ideas. I have delivered such a course of lectures at Canterbury over a number of years and this book has grown out of the lecture notes I have prepared. Although there is no shortage of excellent accounts of the subject of this book (many are listed in the bibliography at the end) the great majority of these are intended for study by post-graduate students who are specializing in this or related fields and are not suitable as adjuncts to an introductory course of lectures at the undergraduate level. I hope, therefore, that there is a place for this book and that university teachers who are responsible for courses of this type and their students will find it helpful.

The plan of the book is as follows: The basic idea of the covariance of physical laws with respect to transformations between reference frames is introduced in Chapter 1 and is illustrated by showing that the laws of Newtonian mechanics are covariant with respect to transformations between inertial frames. This leads to the special principle of relativity and its verification for electrodynamics by the Michelson-Morley experiment. The details of this crucial experiment are not discussed and its aftermath is only briefly alluded to, since I feel that, though of great historical interest, now that more than half a century has elapsed since it was performed, its significance for the special theory, so strongly supported by numerous other experimental results, is not now so great as formerly. The remainder of the first chapter is devoted to the Lorentz transformation which, from the outset, is treated as a rotation of rectangular Cartesian axes in Minkowski space-time. I have thought it desirable to relieve the undergraduate reader, for whom this book is primarily intended, of the necessity for coping immediately with the full rigours of general tensor calculus and I have therefore treated the Minkowski space-time of the special theory as a Euclidean manifold, with the formalism of which he will already be familiar. The time components of vectors are then, of course, purely imaginary, but this is salutary in one respect for it serves to emphasize the basic physical distinction which exists between space and time measurements and so to check the impression that relativity theory implies that space and time are basically of the same nature. The theory of tensors relative to rectangular Cartesian coordinate frames in an N-dimensional Euclidean space is developed in Chapter 2 and this is made use of in the succeeding two chapters to define the principal vectors and tensors of special relativity physics in the Minkowski space-time continuum. All such vectors and tensors are denoted by bold roman capital letters and their components by the corresponding italic characters with appropriate subscripts (e.g. $\mathbf{F}$, F_i for the 4-force vector and its components). The corresponding 3-vectors defined with respect to the rectangular axes employed by an inertial observer are denoted by bold roman lower case letters and their components by the corresponding italic lower case symbols with subscripts (e.g. $\mathbf{f}, f_i$ for the 3-force vector and its components). All sets of transformation equations relating the components of a 3-vector relative to different inertial frames are consistently obtained from the corresponding 4-vector transformation equations with respect to a change of axes in space-time. The reader will accordingly become very familiar with this important technique. The laws of mechanics are all expressed in covariant four-dimensional form in Chapter 3 and the electrodynamic laws are exhibited in this form in Chapter 4. A student of physics, for whom the general theory of relativity is of less interest, should proceed no further than Chapter 4 except, perhaps, to read sections 45 and 46 in which the approach to the general principle is outlined.

The techniques of general tensor calculus, which are employed when expounding the general theory of relativity, are explained in Chapter 5. The algebra and analysis of tensors and relative tensors are first constructed in an affinely connected space, the special case of a Riemannian space where the affinity is conveniently related to the metric being studied later. The coordinates of a point are here denoted by x^i (rather than x_i), for the obvious reason that the dx^i are the components of a contravariant vector and that I have never found the apologies for employing subscripts at all acceptable. I am certain that this practice magnifies, quite unnecessarily, the difficulties of the beginner.

The reason why a theory, which accepts as basic the general principle of relativity, must necessarily be a theory of gravitation is first explained in Chapter 6. This leads to Einstein's law of gravitation, the Schwarzschild spherically symmetric metric for empty space with a point singularity and the three standard physical tests of the theory. It will, perhaps, prevent confusion if it is here remarked that, the interval ds between two neighbouring points of space-time is defined in such a way that, if (x, y, z) are rectangular Cartesian coordinates relative to an inertial frame falling freely in the gravitational field and t is time in this frame, then

$$ds^2 = dx^2 + dy^2 + dz^2 - c^2 dt^2.$$

For the mode of development chosen in this book, this proves to be the most convenient definition of ds.

When giving consideration to the manner in which this material should be presented to the reader, I have greatly profited by referring to the original memoirs of the subject and to the books listed in the bibliography. All these works have influenced my own presentation to some extent, but I have found the accounts by Einstein, Møller and Schrödinger especially stimulating. I am happy to acknowledge my indebtedness to the authors of all books which I have consulted. Also, to my colleague Professor W. R. Andress, who has read the manuscript and made many valuable suggestions, I tender my grateful thanks. As a result of our discussions a number of errors and obscurities have been removed.

Certain of the exercises have been taken from examination papers set at the Universities of Cambridge, London and Liverpool. These have been indicated as follows: M.T. (Mathematics Tripos), L.U. (London) and Li. U. (Liverpool). I am grateful to the authorities concerned for permission to make use of these.

D. F. LAWDEN

Mathematics Department University of Canterbury
Christchurch, N.Z. September 1960

Preface to the Second Edition

A small number of typographical errors have been corrected, but the chief improvement has been the addition of a large number of new exercises, many of which have been taken from examination papers set at the University of Canterbury. There are now 114 of these exercises and it is hoped that these will provide the student with an adequate collection for the purpose of testing his understanding of the text.

D. F. LAWDEN

Mathematics Department University of Canterbury, Christchurch, N.Z. September, 1966.

CHAPTER 1

Special Principle of Relativity. Lorentz Transformations

1. Newton's laws of motion

A proper appreciation of the physical content of Newton's three laws of motion is an essential prerequisite for any study of the special theory of relativity. It will be shown that these laws are in accordance with the fundamental principle upon which the theory is based and thus they will also serve as a convenient introduction to this principle.

The first law states that *any particle which is not subjected to forces moves along a straight line at constant speed.* Since the motion of a particle can only be specified relative to some coordinate frame of reference, this statement has meaning only when the reference frame to be employed when observing the particle's motion has been indicated. Also, since the concept of force has not, at this point, received a definition, it will be necessary to explain how we are to judge when a particle is 'not subjected to forces'. It will be taken as an observed fact that if rectangular axes are taken with their origin at the centre of the sun and these axes do not rotate relative to the most distant objects known to astronomy, viz. the extra-galactic nebulae, then the motions of the neighbouring stars relative to this frame are very nearly uniform. The departure from uniformity can reasonably be accounted for as due to the influence of the stars upon one another and the evidence available suggests very strongly that if the motion of a body in a region infinitely remote from all other bodies could be observed, then its motion would always prove to be uniform relative to our reference frame irrespective of the manner in which the motion was initiated.

We shall accordingly regard the first law as asserting that, in a region of space remote from all other matter and empty save for a single test particle, a reference frame can be defined relative to which the particle will always have a uniform motion. Such a frame will be referred to as an *inertial frame*. An example of such an inertial frame which is conveniently employed when discussing the motions of bodies within the solar system has been described above. However, if S is any inertial frame and $\bar{S}$ is another frame whose axes are always parallel to those of S but whose origin moves with a constant velocity **u** relative to S, then $\bar{S}$ also is inertial. For, if **v**, **v̄** are the velocities of the test particle relative to S, $\bar{S}$ respectively, then

$$\bar{\mathbf{v}} = \mathbf{v} - \mathbf{u} \tag{1.1}$$

and, since **v** is always constant, so is **v̄**. It follows, therefore, that a frame whose origin is at the earth's centre and whose axes do not rotate relative to the stars can, for most practical purposes, be looked upon as an inertial frame, for the motion of the earth relative to the sun is very nearly uniform over periods of time which are normally the subject of dynamical calculations. In fact, since the earth's rotation is slow by ordinary standards, a frame which is fixed in this body can also be treated as approximately inertial and this assumption will only lead to appreciable errors when motions over relatively long periods of time are being investigated, e.g. Foucault's pendulum, long range gunnery calculations.

Having established an inertial frame, if it is found by observation that a particle does not have a uniform motion relative to the frame, the lack of uniformity is attributed to the action of a *force* which is exerted upon the particle by some agency. For example, the orbits of the planets are considered to be curved on account of the force of gravitational attraction exerted upon these bodies by the sun and when a beam of charged particles is observed to be deflected when a bar magnet is brought into the vicinity, this phenomenon is understood to be due to the magnetic forces which are supposed to act upon the particles. If **v** is the particle's velocity relative to the frame at any instant t, its acceleration $\mathbf{a} = d\mathbf{v}/dt$ will be non-zero if the particle's motion is not uniform and this quantity is accordingly a convenient measure of the applied force **f**. We take, therefore,

$$\mathbf{f} \propto \mathbf{a},$$
or
$$\mathbf{f} = m\mathbf{a}, \tag{1.2}$$

where m is a constant of proportionality which depends upon the particle and is termed its *mass*. The definition of the mass of a particle

SPECIAL PRINCIPLE OF RELATIVITY 3

will be given almost immediately when it arises quite naturally out of the third law of motion. Equation (1.2) is essentially a definition of force relative to an inertial frame and is referred to as the *second law of motion*. It is sometimes convenient to employ a non-inertial frame in dynamical calculations, in which case a body which is in uniform motion relative to an inertial frame and is therefore subject to no forces, will nonetheless have an acceleration in the non-inertial frame. By equation (1.2), to this acceleration there corresponds a force, but this will not be attributable to any obvious agency and is therefore usually referred to as a 'fictitious' force. Well-known examples of such forces are the centrifugal and Coriolis forces associated with frames which are in uniform rotation relative to an inertial frame, e.g. a frame rotating with the earth. By introducing such 'fictitious' forces, the second law of motion becomes applicable in all reference frames.

According to the third law of motion, *when two particles P and Q interact so as to influence one another's motion, the force exerted by P on Q is equal to that exerted by Q on P but is in the opposite sense.* Defining the *momentum* of a particle relative to a reference frame as the product of its mass and its velocity, it is proved in elementary textbooks that the second and third laws taken together imply that the sum of the momenta of any two particles involved in a collision is conserved. Thus, if m_1, m_2 are the masses of two such particles and $\mathbf{u}_1$, $\mathbf{u}_2$ are their respective velocities immediately before the collision and $\mathbf{v}_1$, $\mathbf{v}_2$ are their respective velocities immediately afterwards, then

$$m_1 \mathbf{u}_1 + m_2 \mathbf{u}_2 = m_1 \mathbf{v}_1 + m_2 \mathbf{v}_2 \tag{1.3}$$

i.e.
$$\frac{m_2}{m_1}(\mathbf{u}_2 - \mathbf{v}_2) = \mathbf{v}_1 - \mathbf{u}_1. \tag{1.4}$$

This last equation implies that the vectors $\mathbf{u}_2 - \mathbf{v}_2$, $\mathbf{v}_1 - \mathbf{u}_1$ are parallel, a result which has been checked experimentally and which constitutes the physical content of the third law. However, equation (1.4) shows that the third law is also, in part, a specification of how the mass of a particle is to be measured and hence provides a definition for this quantity. For

$$\frac{m_2}{m_1} = \frac{|\mathbf{v}_1 - \mathbf{u}_1|}{|\mathbf{u}_2 - \mathbf{v}_2|} \tag{1.5}$$

and hence the ratio of the masses of two particles can be found from the results of a collision experiment. If, then, one particular particle is chosen to have unit mass (e.g. the standard gramme, pound, etc.), the masses of all other particles can, in principle, be determined by permitting them to collide with this standard and then employing equation (1.5).

2. Covariance of the laws of motion

It has been shown in the previous section that the second and third laws are essentially definitions of the physical quantities force and mass relative to a given reference frame. In this section, we shall examine whether these definitions lead to different results when different inertial frames are employed.

Consider first the definition of mass. If the collision between the particles m_1, m_2 is observed from the inertial frame $\bar{S}$, let $\bar{\mathbf{u}}_1$, $\bar{\mathbf{u}}_2$ be the particle velocities before the collision and $\bar{\mathbf{v}}_1$, $\bar{\mathbf{v}}_2$ the corresponding velocities after the collision. By equation (1.1),

$$\bar{\mathbf{u}}_1 = \mathbf{u}_1 - \mathbf{u}, \text{ etc.} \tag{2.1}$$

and hence

$$\bar{\mathbf{v}}_1 - \bar{\mathbf{u}}_1 = \mathbf{v}_1 - \mathbf{u}_1, \quad \bar{\mathbf{u}}_2 - \bar{\mathbf{v}}_2 = \mathbf{u}_2 - \mathbf{v}_2. \tag{2.2}$$

It follows that if the vectors $\mathbf{v}_1 - \mathbf{u}_1$, $\mathbf{u}_2 - \mathbf{v}_2$ are parallel, so are the vectors $\bar{\mathbf{v}}_1 - \bar{\mathbf{u}}_1$, $\bar{\mathbf{u}}_2 - \bar{\mathbf{v}}_2$ and consequently that, in so far as the third law is experimentally verifiable, it is valid in all inertial frames if it is valid in one. Now let $\bar{m}_1$, $\bar{m}_2$ be the particle masses as measured in $\bar{S}$. Then, by equation (1.5),

$$\frac{\bar{m}_2}{\bar{m}_1} = \frac{|\bar{\mathbf{v}}_1 - \bar{\mathbf{u}}_1|}{|\bar{\mathbf{u}}_2 - \bar{\mathbf{v}}_2|} = \frac{|\mathbf{v}_1 - \mathbf{u}_1|}{|\mathbf{u}_2 - \mathbf{v}_2|} = \frac{m_2}{m_1}. \tag{2.3}$$

But, if the first particle is the unit standard, then $m_1 = \bar{m}_1 = 1$ and hence

$$\bar{m}_2 = m_2, \tag{2.4}$$

i.e. the mass of a particle has the same value in all inertial frames. We can express this by saying that mass is an *invariant* relative to transformations between inertial frames.

By differentiating equation (1.1) with respect to the time t, since $\mathbf{u}$ is constant it is found that

$$\bar{\mathbf{a}} = \mathbf{a}, \tag{2.5}$$

where $\mathbf{a}$, $\bar{\mathbf{a}}$ are the accelerations of a particle relative to S, $\bar{S}$ respectively. Hence, by the second law (1.2), since $\bar{m} = m$, it follows that

$$\bar{\mathbf{f}} = \mathbf{f}, \tag{2.6}$$

i.e. the force acting upon a particle is independent of the inertial frame in which it is measured.

It has therefore been shown that equations (1.2), (1.4) take precisely the same form in the two frames, S, $\bar{S}$, it being understood that mass, acceleration and force are independent of the frame and that velocity is transformed in accordance with equation (1.1). When equations preserve their form upon transformation from one reference frame to another, they are said to be *covariant* with respect to such a transformation. Newton's laws of motion are covariant with respect to a transformation between inertial frames.

3. Special principle of relativity

The special principle of relativity asserts that *all physical laws are covariant with respect to a transformation between inertial frames*. This implies that all observers moving uniformly relative to one another and employing inertial frames will be in agreement concerning the statement of physical laws. No such observer, therefore, can regard himself as being in a special relationship to the universe not shared by any other observer employing an inertial frame; there are no privileged observers. When man believed himself to be at the centre of creation both physically and spiritually, a principle such as that we have just enunciated would have been rejected as absurd. However, the revolution in attitude to our physical environment initiated by Copernicus has proceeded so far that today the principle is accepted as eminently reasonable and very strong evidence contradicting the principle would have to be discovered to disturb it as a foundation upon which theoretical physics is based.

It has been shown already that Newton's laws of motion obey the principle. Let us now transfer our attention to another set of fundamental laws governing non-mechanical phenomena, viz. Maxwell's laws of electrodynamics. These are more complex than the laws of Newton and are most conveniently expressed by the equations

$$\operatorname{curl} \mathbf{E} = -\frac{1}{c}\frac{\partial \mathbf{H}}{\partial t}, \tag{3.1}$$

$$\operatorname{curl} \mathbf{H} = \frac{1}{c}\left(4\pi\mathbf{j} + \frac{\partial \mathbf{E}}{\partial t}\right), \tag{3.2}$$

$$\operatorname{div} \mathbf{E} = 4\pi\rho, \tag{3.3}$$

$$\operatorname{div} \mathbf{H} = 0, \tag{3.4}$$

where $\mathbf{E}$, $\mathbf{H}$ are the electric and magnetic field intensities respectively, $\mathbf{j}$ is the current density, ρ is the charge density and the region of space being considered is assumed to be empty save for the presence of the electric charge. Units have been taken to be Gaussian, so that c is the ratio of the electromagnetic unit of charge to the electrostatic unit ($= 3 \times 10^{10}$ cm/sec). Experiment confirms that these equations are valid when any inertial frame is employed. The most famous such experiment was that carried out by Michelson and Morley, who verified that the velocity of propagation of light waves in any direction is always measured to be c relative to an apparatus stationary on the earth. As is well-known, light has an electromagnetic character and this result is predicted by the equations (3.1)–(3.4). However, the velocity of the earth in its orbit at any time differs from its velocity six months later by twice the orbital velocity, viz. 60 km/sec and thus, by taking measurements of the velocity of light relative to the earth on two days separated by this period of time and showing them to be equal, it is possible to confirm that Maxwell's equations conform to the special principle of relativity. This is effectively what Michelson and Morley did. However, this interpretation of the results of their experiment was not accepted immediately, since it was thought that electromagnetic phenomena were supported by a medium called the *aether* and that Maxwell's equations would prove to be valid only in an inertial frame stationary in this medium, i.e. the special principle of relativity was denied for electromagnetic phenomena. The controversy which ensued is of great historical interest, but will not be recounted in this book. The special principle is now firmly established

SPECIAL PRINCIPLE OF RELATIVITY 7

and is accepted on the grounds that the conclusions which may be deduced from it are everywhere found to be in conformity with experiment and also because it is felt to possess *a priori* a high degree of plausibility. A description of the steps by which it ultimately came to be appreciated that the principle was of quite general application would therefore be superfluous in an introductory text. It is, however, essential for our future development of the theory to understand the prime difficulty preventing an early acceptance of the idea that the electromagnetic laws are in conformity with the special principle.

Consider the two inertial frames S, $\bar{S}$. Suppose that an observer employing S measures the velocity of a light pulse and finds it to be $\mathbf{c}$. If the velocity of the same light pulse is measured by an observer employing the frame $\bar{S}$, let this be $\bar{\mathbf{c}}$. Then, by equation (1.1),

$$\bar{\mathbf{c}} = \mathbf{c} - \mathbf{u} \qquad (3.5)$$

and it is clear that, in general, the magnitudes of the vectors $\bar{\mathbf{c}}$, $\mathbf{c}$ will be different. It appears to follow, therefore, that either Maxwell's equations (3.1)–(3.4) must be modified, or the special principle of relativity abandoned for electromagnetic phenomena. Attempts were made (e.g. by Ritz) to modify Maxwell's equations, but certain consequences of the modified equations could not be confirmed experimentally. Since the special principle was always found to be valid, the only remaining alternative was to reject equation (1.1) and to replace it by another in conformity with the experimental result that the speed of light is the same in all inertial frames. As will be shown in the next section, this can only be done at the expense of a radical revision of our intuitive ideas concerning the nature of space and time and this was very understandably strongly resisted.

4. Lorentz transformations. Minkowski space-time

Let the reference frame S comprise rectangular Cartesian axes $Oxyz$. We shall assume that the coordinates of a point relative to this frame are measured by the usual procedure and employing a measuring scale which is stationary in S (it is necessary to state this precaution, since it will be shown later that the length of a bar is not independent of its motion). It will also be supposed that clocks, stationary relative to S, are distributed throughout space and are all synchronized with a master clock at O. A satisfactory synchronization procedure would be as follows: Warn observers at all clocks that a light source at O will commence radiating at $t = t_0$. When an observer at a point P first receives light from this source, he is to set the clock at P to read $t_0 + OP/c$, i.e. it is assumed that light travels with a speed c relative to S, as found by experiment. The position and time of an event can now be specified relative to S by four coordinates (x, y, z, t), t being the time shown on the clock which is contiguous to the event. We shall often refer to the four numbers (x, y, z, t) as an *event*.

Let $\bar{O}\bar{x}\bar{y}\bar{z}$ be rectangular Cartesian axes determining the frame $\bar{S}$ (to be precise, these are rectangular as seen by an observer stationary in $\bar{S}$) and suppose that clocks at rest relative to this frame are synchronized with a master at $\bar{O}$. Any event can now be fixed relative to $\bar{S}$ by four coordinates $(\bar{x}, \bar{y}, \bar{z}, \bar{t})$, the space coordinates being measured by scales which are at rest in $\bar{S}$ and the time coordinate by the contiguous clock at rest in $\bar{S}$. If (x, y, z, t), $(\bar{x}, \bar{y}, \bar{z}, \bar{t})$ relate to the same event, in this section we are concerned to find the equations relating these corresponding coordinates.

The possibility that the length of a scale and the rate of a clock may be affected by uniform motion relative to a reference frame was ignored in early physical theories. Velocity measurements were agreed to be dependent upon the reference frame, but lengths and time measurements were thought to be absolute. We shall make no such assumption, but will choose the equations relating the coordinates of an event in the two frames to be of such a form that (i) a particle which has uniform motion relative to one frame, has uniform motion relative to the other and (ii) the velocity of propagation of light is the same constant c in both frames. Unless (i) is true, Newton's first law must be abandoned and, with it, the very concept of an inertial frame. Experimental results force us to accept (ii).

To comply with requirement (i), we shall assume that each of the coordinates $(\bar{x}, \bar{y}, \bar{z}, \bar{t})$ is a linear function of the coordinates (x, y, z, t). The inverse relationship is then of the same type. A particle moving uniformly in S with velocity (v_x, v_y, v_z) will have space coordinates (x, y, z) such that

$$x = x_0 + v_x t, \quad y = y_0 + v_y t, \quad z = z_0 + v_z t. \tag{4.1}$$

SPECIAL PRINCIPLE OF RELATIVITY

If linear expressions in the coordinates $(\bar{x}, \bar{y}, \bar{z}, \bar{t})$ are now substituted for (x, y, z, t), it will be found on solving for $(\bar{x}, \bar{y}, \bar{z})$ that these quantities are linear in $\bar{t}$ and hence that the particle's motion is uniform relative to $\bar{S}$. In fact, it may be proved that only a linear transformation can satisfy the requirement (i).

Now suppose that at the instant $t = t_0$ a light source situated at the point P_0 (x_0, y_0, z_0) in S radiates a pulse of short duration. At any later instant t, the wavefront will occupy the sphere whose centre is P_0 and radius $c(t - t_0)$. This has equation

$$(x - x_0)^2 + (y - y_0)^2 + (z - x_0)^2 = c^2(t - t_0)^2. \qquad (4.2)$$

Let $(\bar{x}_0, \bar{y}_0, \bar{z}_0)$ be the coordinates of the light source as observed from $\bar{S}$ at the instant $\bar{t} = \bar{t}_0$ the short pulse is radiated. At any later instant $\bar{t}$, in accordance with requirement (ii), the wavefront must also appear from $\bar{S}$ to occupy a sphere of radius $c(\bar{t} - \bar{t}_0)$ and centre $(\bar{x}_0, \bar{y}_0, \bar{z}_0)$. This has equation

$$(\bar{x} - \bar{x}_0)^2 + (\bar{y} - \bar{y}_0)^2 + (\bar{z} - \bar{z}_0)^2 = c^2(\bar{t} - \bar{t}_0)^2. \qquad (4.3)$$

Equations (4.2), (4.3) describe the same set of events in languages appropriate to $S, \bar{S}$ respectively. It follows that the equations relating the coordinates $(x, y, z, t), (\bar{x}, \bar{y}, \bar{z}, \bar{t})$ must be so chosen that, upon substitution for the 'barred' quantities appearing in equation (4.3) the appropriate linear expressions in the 'unbarred' quantities, equation (4.2) results.

A mathematical device due to Minkowski will now be employed. We shall replace the time coordinate t of any event observed in S by a purely imaginary coordinate $x_4 = ict$ $(i = \sqrt{-1})$. The space coordinates (x, y, z) of the event will be replaced by (x_1, x_2, x_3) so that

$$x = x_1, \quad y = x_2, \quad z = x_3, \quad ict = x_4 \qquad (4.4)$$

and any event is then determined by four coordinates x_i ($i = 1, 2, 3, 4$). A similar transformation to coordinates $\bar{x}_i$ will be carried out in $\bar{S}$. Equations (4.2), (4.3) can then be written

$$\sum_{i=1}^{4} (x_i - x_{i0})^2 = 0, \qquad (4.5)$$

$$\sum_{=1}^{4} (\bar{x} - \bar{x}_{i0})^2 = 0. \qquad (4.6)$$

The $\bar{x}_i$ are to be linear functions of the x_i and such as to transform equation (4.6) into equation (4.5) and hence such that

$$\sum_{i=1}^{4} (\bar{x}_i - \bar{x}_{i0})^2 \to k \sum_{i=1}^{4} (x_i - x_{i0})^2. \tag{4.7}$$

k can only depend upon the relative velocity of S and $\bar{S}$. It is reasonable to assume that the relationship between the two frames is a reciprocal one, so that, when the inverse transformation is made from S to $\bar{S}$, then

$$\sum_{i=1}^{4} (x_i^{\bullet} - x_{i0})^2 \to k \sum_{i=1}^{4} (\bar{x}_i - \bar{x}_{i0})^2. \tag{4.8}$$

But the transformation followed by its inverse must leave any function of the coordinates $\bar{x}_i$ unaltered and hence $k^2 = 1$. In the limit, as the relative motion of S and $\bar{S}$ is reduced to zero, it is clear that $k \to +1$. Hence $k \neq -1$ and we conclude that k is identically unity.

The x_i will now be interpreted as rectangular Cartesian coordinates in a four-dimensional Euclidean space which we shall refer to as $\mathscr{E}_4$. This space is termed *Minkowski space-time*. The left-hand member of equation (4.5) is then the square of the 'distance' between two points having coordinates x_i, x_{i0}. It is now clear that the $\bar{x}_i$ can be interpreted as the coordinates of the point x_i referred to some other rectangular Cartesian axes in $\mathscr{E}_4$. For such an interpretation will certainly enable us to satisfy the requirement (4.7) (with $k = 1$). Also, the x_i, $\bar{x}_i$ will then be related by equations of the form

$$\bar{x}_i = \sum_{j=1}^{4} a_{ij} x_j + b_i, \tag{4.9}$$

where $i = 1, 2, 3, 4$ and the a_{ij}, b_i are constants and this relationship is linear. The b_i are the coordinates of the origin of the first set of rectangular axes relative to the second set. The a_{ij} will be shown to satisfy certain identities in Chapter 2 (equations (8.14), (8.15)). It is proved in algebra texts that the relationship between the x_i and $\bar{x}_i$ must be of the form we are assuming, if it is (i) linear and (ii) such as to satisfy the requirement (4.7).

Changing back from the x_i, $\bar{x}_i$ to the original coordinates of an event by equations (4.4), the equations (4.9) provide a means of relating space and time measurements in S with the corresponding

SPECIAL PRINCIPLE OF RELATIVITY

measurements in $\bar{S}$. Subject to certain provisos (e.g. an event which has real coordinates in S, must have real coordinates in $\bar{S}$), this transformation will be referred to as the *General Lorentz Transformation*.

5. The special Lorentz transformation

We shall now investigate the special Lorentz transformation obtained by supposing that the $\bar{x}_i$-axes in $\mathscr{E}_4$ are obtained from the x_i-axes by a rotation through an angle α parallel to the $x_1 x_4$-plane. The origin and

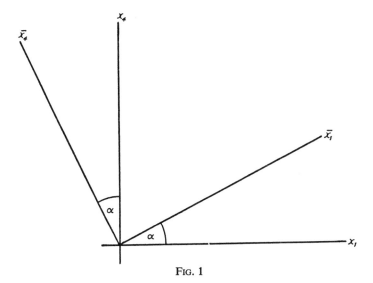

Fig. 1

the x_2, x_3-axes are unaffected by the rotation and it will be clear after consideration of Fig. 1 therefore that

$$\left.\begin{aligned}\bar{x}_1 &= x_1 \cos\alpha + x_4 \sin\alpha, & \bar{x}_2 &= x_2,\\ \bar{x}_4 &= -x_1 \sin\alpha + x_4 \cos\alpha, & \bar{x}_3 &= x_3.\end{aligned}\right\} \quad (5.1)$$

Employing equations (4.4), these transformation equations may be written

$$\left.\begin{aligned}\bar{x} &= x\cos\alpha + ict\sin\alpha, & \bar{y} &= y,\\ ic\bar{t} &= -x\sin\alpha + ict\cos\alpha, & \bar{z} &= z.\end{aligned}\right\} \quad (5.2)$$

To interpret the equations (5.2), consider a plane which is stationary relative to the $\bar{S}$ frame and has equation

$$\bar{a}\bar{x}+\bar{b}\bar{y}+\bar{c}\bar{z}+\bar{d} = 0 \tag{5.3}$$

for all $\bar{t}$. Its equation relative to the S frame will be

$$(\bar{a}\cos\alpha)x+\bar{b}y+\bar{c}z+\bar{d}+ict\bar{a}\sin\alpha = 0, \tag{5.4}$$

at any fixed instant t. In particular, if $\bar{a} = \bar{b} = \bar{d} = 0$, this is the coordinate plane $\bar{O}\bar{x}\bar{y}$ and its equation relative to S is $z = 0$, i.e. it is the

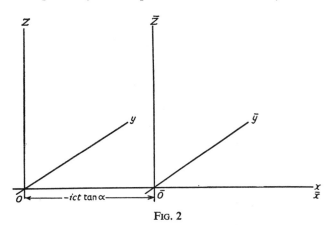

Fig. 2

plane Oxy. Again, if $\bar{b} = \bar{c} = \bar{d} = 0$, the plane is $\bar{O}\bar{y}\bar{z}$ and its equation in S is

$$x = -ict\tan\alpha, \tag{5.5}$$

i.e. it is a plane parallel to Oyz displaced a distance $-ict\tan\alpha$ along Ox. Finally, if $\bar{a} = \bar{c} = \bar{d} = 0$, the plane is $\bar{O}\bar{z}\bar{x}$ and its equation with respect to S is $y = 0$, i.e. it is the plane Ozx. We conclude, therefore, that the Lorentz transformation equations (5.2) correspond to the particular case when the coordinate planes comprising $\bar{S}$ are obtained from those comprising S at any instant t by a translation along Ox a distance $-ict\tan\alpha$ (Fig. 2). Thus, if u is the speed of translation of $\bar{S}$ relative to S

$$u = -ic\tan\alpha. \tag{5.6}$$

It should also be noted that the events

$$x = y = z = t = 0, \quad \bar{x} = \bar{y} = \bar{z} = \bar{t} = 0$$

correspond and hence that, at the instant O and $\bar{O}$ coincide, the S and $\bar{S}$ clocks at these points are supposed set to have zero readings; all other clocks are then synchronized with these.

Equation (5.6) indicates that α is imaginary and is directly related to the speed of translation. We have $\tan\alpha = iu/c$ and hence

$$\cos\alpha = \frac{1}{\sqrt{(1-u^2/c^2)}}, \quad \sin\alpha = \frac{(iu/c)}{\sqrt{(1-u^2/c^2)}}. \quad (5.7)$$

Substituting in the equations (5.2), the special Lorentz transformation is obtained in its final form, viz.

$$\left.\begin{array}{l} \bar{x} = \dfrac{x-ut}{\sqrt{(1-u^2/c^2)}}, \quad \bar{y} = y, \\[2mm] \bar{t} = \dfrac{t-(ux/c^2)}{\sqrt{(1-u^2/c^2)}}, \quad \bar{z} = z. \end{array}\right\} \quad (5.8)$$

If u is small by comparison with c, as is generally the case, these equations may evidently be approximated by the equations

$$\left.\begin{array}{ll} \bar{x} = x-ut, & \bar{y} = y, \\ \bar{t} = t, & \bar{z} = z. \end{array}\right\} \quad (5.9)$$

This set of equations, called the *Special Galilean Transformation* equations, is, of course, the set which was assumed to relate space and time measurements in the two frames in classical physical theory. However, the equation $\bar{t} = t$ was rarely stated explicitly, since it was taken as self-evident that time measurements were absolute, i.e. quite independent of the observer. It appears from equations (5.8) that this view of the nature of time can no longer be maintained and that, in fact, time and space measurements are related, as is shown by the dependence of $\bar{t}$ upon both t and x. This revolutionary idea is also suggested by the manner in which the special Lorentz transformation has been derived, viz. by a rotation of axes in a manifold which has both spacelike and timelike characteristics. However, this does not imply that space and time are now to be regarded as basically similar physical quantities, for it has only been possible to place the time coordinate on the same footing as the space coordinates in $\mathscr{E}_4$ by multiplying the former by i. Since x_4 must always be imaginary, whereas x_1, x_2, x_3 are real, the fundamentally different nature of space and time measurements is still maintained in the new theory.

If $u > c$, both $\bar{x}$ and $\bar{t}$ as given by equations (5.8) are imaginary. We conclude that no observer can possess a velocity greater than that of light relative to any other observer.

If equations (5.8) are solved for (x, y, z, t) in terms of $(\bar{x}, \bar{y}, \bar{z}, \bar{t})$, it will be found that the inverse transformation is identical with the original transformation, except that the sign of u is reversed. This also follows from the fact that the inverse transformation corresponds to a rotation of axes through an angle $-\alpha$ in space-time. Thus, the frame S has velocity $-\mathbf{u}$ when observed from $\bar{S}$.

6. Fitzgerald contraction. Time dilation

In the next two sections, we shall explore some of the more elementary physical consequences of the transformation equations (5.8).

Consider first a rigid rod stationary in $\bar{S}$ and lying along the $\bar{x}$-axis. Let $\bar{x} = \bar{x}_1$, $\bar{x} = \bar{x}_2$ at the two ends of the bar so that its length as measured in $\bar{S}$ is given by

$$\bar{l} = \bar{x}_2 - \bar{x}_1. \tag{6.1}$$

At the instant t in S, suppose these ends occupy the positions $x = x_1$, $x = x_2$. Then, by equations (5.8),

$$\bar{x}_1 = \frac{x_1 - ut}{\sqrt{(1 - u^2/c^2)}}, \quad \bar{x}_2 = \frac{x_2 - ut}{\sqrt{(1 - u^2/c^2)}}. \tag{6.2}$$

But $x_2 - x_1 = l$ is the length of the bar as measured in S and it follows by subtraction of equations (6.3) that

$$l = \bar{l}\sqrt{(1 - u^2/c^2)}. \tag{6.3}$$

The length of a bar accordingly suffers contraction when it is moved longitudinally relative to an inertial frame. This is the *Fitzgerald contraction*.

SPECIAL PRINCIPLE OF RELATIVITY   15

This contraction is not to be thought of as the physical reaction of the rod to its motion and as belonging to the same category of physical effects as the contraction of a metal rod when it is cooled. It is due to a changed relationship between the rod and the instruments measuring its length. $\bar{l}$ is a measurement carried out by scales which are stationary relative to the bar, whereas l is the result of a measuring operation with scales which are moving with respect to the bar. Also, the first operation can be carried out without the assistance of a clock, but the second operation involves simultaneous observation of the two ends of the bar and hence clocks must be employed. In classical physics, it was assumed that these two measurement procedures would yield the same result, since it was supposed that a rigid bar possessed intrinsically an attribute called its length and that this could in no way be affected by the procedure employed to measure it. It is now understood that length, like every other physical quantity, is *defined* by the procedure employed for its measurement and that it possesses no meaning apart from being the result of this procedure. From this point of view, it is not surprising that, when the procedure must be altered to suit the circumstances, the result will also be changed. It may assist the reader to adopt the modern view of the Fitzgerald contraction if we remark that the length of the rod considered above can be altered at any instant by simply changing our minds and commencing to employ the S frame rather than the $\bar{S}$ frame. Clearly, such a change of mathematical description can have no physical consequences.

Now consider two events which have coordinates (x_1, y_1, z_1, t), (x_2, y_2, z_2, t) in S and hence occur simultaneously at different points. These events are not necessarily simultaneous in $\bar{S}$. For, if $\bar{t} = \bar{t}_1$ at the first mentioned event and $\bar{t} = \bar{t}_2$ at the second, then

$$\bar{t}_2 - \bar{t}_1 = \frac{u}{c^2}(x_1 - x_2)/\sqrt{(1 - u^2/c^2)} \tag{6.4}$$

and $\bar{t}_2 \neq \bar{t}_1$, unless $x_1 = x_2$. The concept of simultaneity is accordingly, also, a relative one and has no absolute meaning as was previously thought.

The registration by the clock moving with $\bar{O}$ of the times $\bar{t}_1, \bar{t}_2$, constitutes two events having coordinates $(0, 0, 0, \bar{t}_1)$, $(0, 0, 0, \bar{t}_2)$ respectively in $\bar{S}$. Employing the inverse transformation to (5.8), it follows that the times t_1, t_2 of these events as measured in S are given by

$$t_1 = \bar{t}_1/\sqrt{(1-u^2/c^2)}, \quad t_2 = \bar{t}_2/\sqrt{(1-u^2/c^2)}, \tag{6.5}$$

and hence that

$$\bar{t}_1 - \bar{t}_2 = (t_1 - t_2)\sqrt{(1-u^2/c^2)}. \tag{6.6}$$

This equation shows that the clock moving with $\bar{O}$ will appear from S to have its rate reduced by a factor $\sqrt{(1-u^2/c^2)}$. This is the *time dilation* effect.

Since any cyclic physical process, i.e. one which returns to some initial state after the lapse of a period of time, can be employed as a clock, the result just obtained implies that all physical processes will evolve more slowly when observed from a frame relative to which they are moving. Thus, the rate of decay of radioactive particles present in cosmic rays and moving with high velocities relative to the earth, has been observed to be reduced by exactly the factor predicted by equation (6.6).

It may also be deduced that, if a human passenger were to be launched from the earth in a rocket which attained a speed approaching that of light and after proceeding to a great distance returned to the earth with the same high speed, suitable observations made from the earth would indicate that all physical processes occurring within the rocket, including the metabolic and physiological processes taking place inside the passenger's body, would suffer a retardation. Since all physical processes would be affected equally, the passenger would be unaware of this effect. Nonetheless, upon return to the earth he would find that his estimate of the duration of the flight was less than the terrestrial estimate. It may be objected that the passenger is entitled to regard himself as having been at rest and the earth as having suffered the displacement and therefore that the terrestrial estimate should be less than his own. This is the *clock paradox*. The paradox is resolved by observing that a frame moving with the rocket is subject to an acceleration relative to an inertial frame and consequently cannot be treated as inertial. The results of special relativity only apply to inertial frames and the rocket passenger is accordingly not entitled to make use of them in his own frame. As will be shown later, the methods of general relativity theory are applicable in any frame and it may be proved

SPECIAL PRINCIPLE OF RELATIVITY   17

that, if the passenger employs these methods, his calculations will yield results in agreement with those obtained by the terrestrial observer.

7. Spacelike and timelike intervals. Light cone

We have proved in section 4 that if x_i, x_{i0} are the coordinates in Minkowski space-time of two events, then

$$\sum_{i=1}^{4} (x_i - x_{i0})^2 \tag{7.1}$$

is invariant, i.e. has the same value for all observers employing inertial frames and thus rectangular axes in space-time. Reverting by equations (4.4) to the ordinary space and time coordinates employed in an inertial frame, it follows that

$$(x-x_0)^2 + (y-y_0)^2 + (z-z_0)^2 - c^2(t-t_0)^2, \tag{7.2}$$

is invariant for all inertial observers.

Thus, if (x,y,z,t), (x_0,y_0,z_0,t_0) are the coordinates of two events relative to any inertial frame S and we define the *proper time interval* τ between the events by the equation

$$\tau^2 = (t-t_0)^2 - \frac{1}{c^2}\{(x-x_0)^2 + (y-y_0)^2 + (z-z_0)^2\}, \tag{7.3}$$

then τ is an invariant for the two events. Two observers employing different inertial frames may attribute different coordinates to the events, but they will be in agreement concerning the value of τ.

Denoting the time interval between the events by Δt and the distance between them by Δd, both relative to the same frame S and positive, it follows from equation (7.3) that

$$\tau^2 = \Delta t^2 - \frac{1}{c^2}\Delta d^2. \tag{7.4}$$

Suppose that a new inertial frame $\bar{S}$ is now defined, moving in the direction of the line joining the events with speed $\Delta d/\Delta t$. This will only be possible if $\Delta d/\Delta t < c$. Relative to this frame the events will occur at the same point and hence $\overline{\Delta d} = 0$. By equation (7.4), therefore,

$$\tau^2 = \overline{\Delta t}^2, \tag{7.5}$$

i.e. the proper time interval between two events is the ordinary time interval measured in a frame (if such exists) in which the events occur at the same space point. In this case, it is clear that $\tau^2 > 0$ and the proper time interval between the events is said to be *timelike*.

Suppose, if possible, that a frame $\bar{S}$ can be chosen relative to which the events are simultaneous. In this frame $\overline{\Delta t} = 0$ and

$$\tau^2 = -\frac{1}{c^2}\overline{\Delta d}^2. \tag{7.6}$$

Thus $\tau^2 < 0$, and, in any frame, $\Delta d/\Delta t > c$. τ is then purely imaginary and the interval between the events is said to be *spacelike*.

If the interval is timelike, $\Delta d/\Delta t < c$ and it is possible for a material body to be present at both events. On the other hand, if the interval is spacelike, $\Delta d/\Delta t > c$ and it is not possible for such a body to be present at both events. The intermediate case is when $\Delta d/\Delta t = c$ and then $\tau = 0$. Only a light pulse can be present at both events. It also follows that the proper time interval between the transmission and reception of a light signal is zero.

We shall now represent the event (x, y, z, t) by a point having these coordinates in a four-dimensional space. This space is also often referred to as Minkowski space-time but, unlike the space-time continuum introduced in section 4, it is not Euclidean. However, this representation has the advantage that the coordinates all take real values and it is therefore more satisfactory when diagrams are to be drawn. Suppose a particle is at the origin O of S at $t = 0$ and commences to move along Ox with constant speed u. Its y- and z-coordinates will always be zero and the representation of its motion in space-time will be confined to the xt-plane. In this plane, its motion will appear as the straight line QP, Q being the point $x = y = z = t = 0$ (Fig. 3). QP is called the *world-line* of the particle. If $\angle PQt = \theta$, $\tan\theta = u$. But $|u| \leq c$ and hence the world-line of the particle must lie in the sector AQB, where $\angle AQB = 2\alpha$ and $\tan\alpha = c$. Similarly, the world-line of a particle which arrives at O at $t = 0$ after moving along Ox, must lie in the sector $A'QB'$. It follows that any event in either of these sectors must be separated from the event Q by a timelike interval, since a particle can be present at both events. Events in the sectors AQB', $A'QB$ are separated from Q by spacelike intervals. $A'A$, $B'B$ are the world-lines of light signals passing through O at $t = 0$ and

SPECIAL PRINCIPLE OF RELATIVITY 19

being propagated in the directions of the positive and negative *x*-axis respectively.

For any event in AQB, $t > 0$, i.e. it is in the future with respect to the

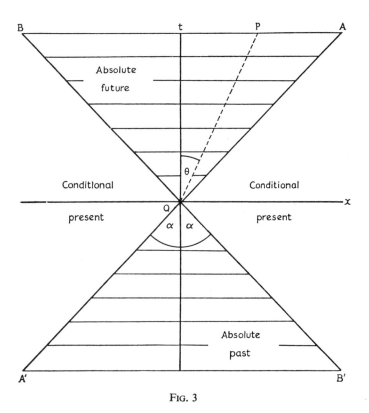

Fig. 3

event Q when the frame S is being employed. However, by no choice of frame can it be made simultaneous with Q, for this would imply a spacelike interval. *A fortiori*, in no frame can it occur prior to Q. The sector AQB accordingly contains events which are in the *absolute future* with respect to the event Q. Similarly all events in the sector $A'QB'$ are in the *absolute past* with respect to Q. On the other hand, events lying in the sectors AQB', $A'QB$ are separated from Q by spacelike intervals and can all be made simultaneous with Q by proper choice of inertial frame. These events may occur before or after Q depending upon the frame being used. These two sectors define a region of space-time which will be termed the *conditional present*.

Since no physical signal can have a speed greater than c, the world-line of any such signal emanating from Q must lie in the sector AQB. It follows that the event Q can be the physical cause of only those events which are in the absolute future with respect to Q. Similarly, Q can be the effect of only those events in its absolute past. Q cannot be causally related to events in its conditional present.

This state of affairs should be contrasted with the essentially simpler situation of classical physics where there is no upper limit to the signal velocity and AA', BB' coincide along the x-axis. Past and future are then separated by a perfectly precise present in which events all have the time coordinate $t = 0$ for all observers.

In the four-dimensional space $Qxyzt$, the three regions of absolute past, absolute future and conditional present are separated from one another by the hyper-cone

$$x^2 + y^2 + z^2 - c^2 t^2 = 0. \tag{7.7}$$

A light pulse transmitted from Q will have its world-line on this surface, which is accordingly called the *light cone* at Q. Since any arbitrary event can be selected to be Q, any event is the apex of a light cone which separates the space-time continuum in an absolute manner into three distinct regions relative to the event.

Exercises 1

1. A particle of mass m is moving in the plane of axes Oxy under the action of a force **f**. Oxy is an inertial frame. $Ox'y'$ is rotating relative to the inertial frame so that $\angle x'Ox = \psi$ and $\dot\psi = \omega$. (r, θ) are polar coordinates of the particle relative to the rotating frame. If (f_r, f_θ) are the polar components of **f**, (a_r, a_θ) are the polar components of the

particle's acceleration relative to $Ox'y'$, v is the particle's speed relative to this frame and ϕ is the angle its direction of motion makes with the radius vector in this frame, obtain the equations of motion in the form

$$ma_r = f_r + 2m\omega v \sin\phi + mr\omega^2,$$

$$ma_\theta = f_\theta - 2m\omega v \cos\phi - mr\dot\omega.$$

Deduce that the motion relative to the rotating frame is in accordance with the second law if, in addition to **f**, the following forces are also taken to act on the particle: (i) $m\omega^2 r$ radially outwards (the centrifugal force), (ii) $2m\omega v$ at right angles to the direction of motion (the Coriolis force), (iii) $mr\dot\omega$ transversely. (The latter force vanishes if the rotation is uniform.)

2. A bar lies along $\bar{O}\bar{x}$ and is stationary in $\bar{S}$. Show that if the positions of its ends are observed in S at instants which are simultaneous in $\bar{S}$, its length deduced from these observations will be greater than its length in $\bar{S}$ by a factor $(1 - u^2/c^2)^{-1/2}$.

3. Suppose that the bar referred to in Exercise 2 takes a time T to pass a fixed point on the x-axis, T being measured by a clock stationary at the fixed point. Defining the length of the bar in the S-frame to be uT, deduce the Fitzgerald contraction.

4. The measuring rod employed by S will appear from $\bar{S}$ to be shortened by a factor $(1-u^2/c^2)^{1/2}$. Hence, when S measures the length of the bar fixed in $\bar{S}$ he might be expected to obtain the result

$$l = \bar{l}/(1-u^2/c^2)^{1/2}.$$

This contradicts equation (6.3). Resolve the contradiction. [Hint: It will be observed from $\bar{S}$ that S fixes the position of the forward end of the bar first and the position of the rear end a time $u\bar{l}/c^2$ later.]

5. A bar lies stationary along the x-axis of S. Show that the world-lines of the particles of the bar occupy a certain 'band' in the $x_1 x_4$-plane. By measuring the width of this 'band' parallel to the $\bar{x}_1$-axis, deduce the Fitzgerald contraction.

6. Verify that the transformation equations (5.8) are such that

$$\bar{x}^2 + \bar{y}^2 + \bar{z}^2 - c^2 \bar{t}^2 = x^2 + y^2 + z^2 - c^2 t^2.$$

7. Two light pulses are moving in the positive direction along the x-axis of the frame S, the distance between them being d. Show that, as measured in $\bar{S}$, the distance between the pulses is

$$d \bigg/ \sqrt{\left(\frac{c+u}{c-u}\right)}.$$

8. A and B are two points of an inertial frame S a distance d apart. An event occurs at B a time T (relative to clocks in S) after another event occurs at A. Relative to another inertial frame $\bar{S}$, the events are simultaneous. If **AP** is a displacement vector in S representing the velocity of $\bar{S}$ relative to S, prove that P lies in a plane perpendicular to AB, distance $c^2 T/d$ from A.

9. S, $\bar{S}$ are the inertial frames considered in section 5. The length of a moving rod, which remains parallel to the x and $\bar{x}$ axes, is measured as a in the frame S and $\bar{a}$ in the frame $\bar{S}$. By consideration of a Minkowski diagram for the rod, or otherwise, show that the proper length of the rod is

$$\frac{a\bar{a}\beta u/c}{\sqrt{(2\beta a\bar{a} - a^2 - \bar{a}^2)}},$$

where $\beta = (1 - u^2/c^2)^{-1/2}$.

10. If the position vectors $\mathbf{r} = (x, y, z)$, $\bar{\mathbf{r}} = (\bar{x}, \bar{y}, \bar{z})$ of an event as determined by the observers in the parallel inertial frames S, $\bar{S}$ respectively are mapped in the same independent $\mathscr{E}_3$, prove that

$$\bar{\mathbf{r}} = \mathbf{r} + \mathbf{u}\left\{\frac{\mathbf{u}\cdot\mathbf{r}}{u^2}(\beta - 1) + \beta t\right\},$$
$$\bar{t} = \beta(t + \mathbf{u}\cdot\mathbf{r}/c^2),$$

where $\beta = (1 - u^2/c^2)^{-1/2}$ and $\mathbf{u}$ is the velocity of S as measured from $\bar{S}$.

11. The centroid and axis of a right circular cylinder are fixed in the inertial frame S relative to which the cylinder rotates about its axis with uniform velocity ω. Prove that, when observed from $\bar{S}$, the cylinder will appear to be twisted about its axis through an angle $u\omega/c^2$ per unit rest-length of the cylinder.

CHAPTER 2

Orthogonal Transformations. Cartesian Tensors

8. Orthogonal transformations

In section 4 events have been represented by points in a space $\mathscr{E}_4$. The resulting distribution of points was described in terms of their coordinates relative to a set of rectangular Cartesian axes. Each such set of axes was shown to correspond to an observer employing a rectangular Cartesian inertial frame in ordinary $\mathscr{E}_3$-space and clocks which are stationary in this frame. In this representation, the descriptions of physical phenomena given by two such inertial observers are related by a transformation in $\mathscr{E}_4$ from one set of rectangular axes to another. Such a transformation has been given at equation (4.9) and is called an *orthogonal* transformation. In general, if x_i, $\bar{x}_i$ ($i = 1, 2, \ldots, N$) are two sets of N quantities which are related by a linear transformation

$$\bar{x}_i = \sum_{j=1}^{N} a_{ij} x_j + b_i, \tag{8.1}$$

and, if the coefficients a_{ij} of this transformation are such that

$$\sum_{i=1}^{N} (\bar{x}_i - \bar{y}_i)^2 = \sum_{i=1}^{N} (x_i - y_i)^2 \tag{8.2}$$

is an identity for all corresponding sets x_i, $\bar{x}_i$ and y_i, $\bar{y}_i$, then the transformation is said to be orthogonal. It is clear that the x_i, $\bar{x}_i$ may be thought of as the coordinates of a point in $\mathscr{E}_N$ referred to two different sets of rectangular Cartesian axes and then equation (8.2) states that the square of the distance between two points is an invariant, independent of the Cartesian frame.

Writing $z_i = x_i - y_i$, $\bar{z}_i = \bar{x}_i - \bar{y}_i$, it follows from equation (8.1) that

$$\bar{z}_i = \sum_{j=1}^{N} a_{ij} z_j. \tag{8.3}$$

Let z denote the column matrix with elements z_i, $\bar{z}$ the column matrix with elements $\bar{z}_i$ and A the $N \times N$ matrix with elements a_{ij}. Then the set of equations (8.3) is equivalent to the matrix equation

$$\bar{z} = Az. \tag{8.4}$$

Also, if z' is the transpose of z,

$$z'z = \sum_{i=1}^{N} z_i^2 \tag{8.5}$$

and thus the identity (8.2) may be written

$$\bar{z}'\bar{z} = z'z. \tag{8.6}$$

But, from equation (8.4),

$$\bar{z}' = z'A'. \tag{8.7}$$

Substituting in the left-hand member of equation (8.6) from equations (8.4), (8.7), it will be found that

$$z'A'Az = z'z. \tag{8.8}$$

This can only be true for all z if

$$A'A = I, \tag{8.9}$$

where I is the unit $N \times N$ matrix.

Taking determinants of both members of the matrix equation (8.9), we find that $|A|^2 = 1$ and hence

$$|A| = \pm 1. \tag{8.10}$$

A is accordingly regular. Let A^{-1} be its inverse. Multiplication on the right by A^{-1} of both members of equation (8.9) then yields

$$A' = A^{-1}. \tag{8.11}$$

It now follows that

$$AA' = AA^{-1} = I. \tag{8.12}$$

Let δ_{ij} be the ij^{th} element of I, so that

$$\left.\begin{array}{ll} \delta_{ij} = 1, & i = j, \\ = 0, & i \neq j. \end{array}\right\} \tag{8.13}$$

The symbols δ_{ij} are referred to as the *Kronecker deltas*. Equations (8.9), (8.12) are now seen to be equivalent to

$$\sum_{i=1}^{N} a_{ij} a_{ik} = \delta_{jk}, \tag{8.14}$$

$$\sum_{i=1}^{N} a_{ji} a_{ki} = \delta_{jk} \tag{8.15}$$

respectively. These conditions are necessarily satisfied by the coefficients a_{ij} of the transformation (8.1) if it is orthogonal. Conversely, if either of these conditions is satisfied, it is easy to prove that equation (8.6) follows and hence that the transformation is orthogonal.

9. Repeated index summation convention

At this point it is convenient to introduce a notation which will greatly abbreviate future manipulative work. It will be understood that, wherever in any term of an expression a literal index occurs twice, this term is to be summed over all possible values of the index. For example, we shall abbreviate by writing

$$\sum_{r=1}^{N} a_r b_r = a_r b_r. \tag{9.1}$$

The index must be a literal one and we shall further stipulate that it must be a small letter. Thus $a_2 b_2$, $a_N b_N$ are individual terms of the expression $a_r b_r$, and no summation is intended in these cases.

Employing this convention, equations (8.14) and (8.15) can be written

$$a_{ij} a_{ik} = \delta_{jk}, \quad a_{ji} a_{ki} = \delta_{jk} \tag{9.2}$$

respectively. Again, with $z_i = x_i - y_i$, equation (8.2) may be written

$$\bar{z}_i \bar{z}_i = z_i z_i. \tag{9.3}$$

More than one index may be repeated in the same term, in which case more than one summation is intended. Thus

$$a_{ij} b_{jk} c_k = \sum_{j=1}^{N} \sum_{k=1}^{N} a_{ij} b_{jk} c_k. \tag{9.4}$$

It is permissible to replace a repeated index by any other small letter, provided the replacement index does not occur elsewhere in the same term. Thus

$$a_i b_i = a_j b_j = a_k b_k, \tag{9.5}$$

but

$$a_{ij} a_{ik} \neq a_{jj} a_{jk}, \tag{9.6}$$

irrespective of whether the right-hand member is summed with respect to j or not. A repeated index shares this property with the variable of integration in a definite integral. Thus

$$\int_a^b f(x)\,dx = \int_a^b f(y)\,dy. \tag{9.7}$$

A repeated index is accordingly referred to as a *dummy index*. Any other index will be called a *free index*.

It will be assumed, in future, that any equation remains true when the free indices assume all possible values. Thus, equations (9.2) are true for all $j = 1, 2, \ldots, N$ and all $k = 1, 2, \ldots, N$. It is clear that the free indices on the two sides of an equation will be identical.

The reader should note carefully the identity

$$\delta_{ij} a_j = a_i, \tag{9.8}$$

for it will be of frequent application. δ_{ij} is often called a *substitution operator*, since when it multiplies a symbol such as a_j, its effect is to replace the index j by i.

10. Rectangular Cartesian tensors

Let x_i, y_i be rectangular Cartesian coordinates of two points Q, P respectively in $\mathscr{E}_N$. Writing $z_i = x_i - y_i$, the z_i are termed the *components* of the *displacement vector* **PQ** relative to the axes being used. If $\bar{x}_i$, $\bar{y}_i$ are the coordinates of Q, P with respect to another set of rectangular axes, the new coordinates will be related to the old by the transformation equations (8.1). Then, if $\bar{z}_i$ are the components of **PQ** in the new frame, it follows (equation (8.3)) that

$$\bar{z}_i = a_{ij} z_j. \tag{10.1}$$

Any set of N quantities which take the values A_i when the first coordinate frame is being employed and which transform in the same manner as the z_i when referred to a new coordinate frame, i.e. are such that

$$\bar{A}_i = a_{ij} A_j, \quad (10.2)$$

are said to be the components of a *vector* in $\mathscr{E}_N$ relative to rectangular Cartesian reference frames. We shall frequently abbreviate 'the vector whose components are A_i' to 'the vector A_i'. We shall also denote the vector by **A**.

If A_i, B_i are two vectors, consider the N^2 quantities $A_i B_j$. Upon transformation of axes, these quantities transform thus:

$$\bar{A}_i \bar{B}_j = a_{ik} a_{jl} A_k B_l. \quad (10.3)$$

Any set of N^2 quantities C_{ij} which transform in this manner, i.e. which are such that

$$\bar{C}_{ij} = a_{ik} a_{jl} C_{kl}, \quad (10.4)$$

are said to be the components of a *tensor* of the second *rank*. We shall speak of 'the tensor C_{ij}'. Such a tensor is not, necessarily, representable as the product of two vectors.

A set of N^3 quantities D_{ijk} which transform in the same manner as the product of three vectors $A_i B_j C_k$, form a tensor of the third rank. The transformation law is

$$\bar{D}_{ijk} = a_{il} a_{jm} a_{kn} D_{lmn}. \quad (10.5)$$

The generalization to a tensor of any rank should now be obvious. Vectors are, of course, tensors of the first rank.

If A_{ij}, B_{ij} are tensors, the sums $A_{ij} + B_{ij}$ are N^2 quantities which transform according to the same law as the A_{ij} and B_{ij}. The sum of two tensors of the second rank is accordingly also a tensor of this rank. This result can be generalized immediately to the sum of any two tensors of identical rank. Similarly, the difference of two tensors of the same rank is also a tensor.

Our method of introducing a tensor implies that the product of any number of vectors is a tensor. Quite generally, if $A_{ij\ldots}$, $B_{ij\ldots}$ are tensors of any ranks (which may be different), then the product $A_{ij\ldots} B_{kl\ldots}$ is a tensor whose rank is the sum of the ranks of the two factors. The reader should prove this formally for a product such as

$A_{ij}B_{klm}$, by writing down the transformation equations. (N.B. the indices in the two factors must be kept distinct, for otherwise a summation is implied and this complicates matters; see section 12.)

The components of a tensor may be chosen arbitrarily relative to any one set of axes. The components of the tensor relative to any other set are then fixed by the transformation equations. Consider the tensor of the second rank whose components relative to the x_i-axes are the Kronecker deltas δ_{ij}. In the $\bar{x}_i$-frame, the components are

$$\bar{\delta}_{ij} = a_{ik}a_{jl}\delta_{kl} = a_{ik}a_{jk} = \delta_{ij}, \qquad (10.6)$$

by equations (9.2). Thus this tensor has the same components relative to all sets of axes. It is termed the *fundamental tensor* of the second rank.

If, to take the particular case of a third rank tensor as an example,

$$A_{ijk} = A_{jik} \qquad (10.7)$$

for all values of i, j, k, A_{ijk} is said to be *symmetric* with respect to its indices i, j. Symmetry may be with respect to any pair of indices. If A_{ijk} is a tensor, its property of symmetry with respect to two indices is preserved upon transformation, for

$$\bar{A}_{jik} = a_{jl}a_{im}a_{kn}A_{lmn},$$
$$= a_{im}a_{jl}a_{kn}A_{mln},$$
$$= \bar{A}_{ijk}, \qquad (10.8)$$

where, in the second line, we have rearranged and put $A_{lmn} = A_{mln}$. Unless a property is preserved upon transformation, it will be of little importance to us, for we shall later employ tensors to express relationships which are valid for all observers and a chance relationship, true in one frame alone, will be of no fundamental significance.

Similarly, if

$$A_{ijk} = -A_{jik} \qquad (10.9)$$

for all values of i, j, k, A_{ijk} is said to be *skew-symmetric* or *antisymmetric* with respect to its first two indices. This property also is preserved upon transformation. Since $A_{11k} = -A_{11k}$, $A_{11k} = 0$. All components of A_{ijk} with the first two indices the same are clearly zero.

ORTHOGONAL TRANSFORMATIONS 29

A tensor whose components are all zero in one frame, has zero components in every frame. A corollary to this result is that if $A_{ij...}$, $B_{ij...}$ are two tensors of the same rank whose corresponding components are equal in one frame, then they are equal in every frame. This follows because $A_{ij...} - B_{ij...}$ is a tensor whose components are all zero in the first frame and hence in every frame. Thus, a *tensor equation*

$$A_{ij...} = B_{ij...} \tag{10.10}$$

is valid for all choices of axes. This explains the importance of tensors for our purpose. By expressing a physical law as a tensor equation, we shall guarantee its covariance with respect to a change of inertial frame.

11. Invariants. Gradients. Derivatives of tensors

Suppose that V is a quantity which is unaffected by any change of axes. Then V is called a *scalar invariant* or simply an *invariant*. Its transformation equation is simply

$$\bar{V} = V. \tag{11.1}$$

As will be proved later (section 21), the charge of an electron is independent of the inertial frame from which it is measured and is, therefore, the type of quantity we are considering.

If a value of V is associated with each point of a region of $\mathscr{E}_N$, an *invariant field* is defined over this region. In this case V will be a function of the coordinates x_i. Upon transformation to new axes, V will be expressed in terms of the new coordinates $\bar{x}_i$; when so expressed, it is denoted by $\bar{V}$. Thus

$$\bar{V}(\bar{x}_1, \bar{x}_2, ..., \bar{x}_N) = V(x_1, x_2, ..., x_N) \tag{11.2}$$

is an identity. The reader should, perhaps, be warned that $\bar{V}$ is not, necessarily, the same function of the $\bar{x}_i$ that V is of the x_i.

If A_{ij} is a tensor, it is obvious that VA_{ij} is also a tensor of the second rank. It is therefore convenient to regard an invariant as a tensor of zero rank.

Consider the N partial derivatives $\partial V/\partial x_i$. These transform as a vector. To prove this it will be necessary to examine the transformation inverse to (8.1). In the matrix notation of section 8, this may be written

$$x = A^{-1}(\bar{x}-b) = A'(\bar{x}-b), \tag{11.3}$$

having made use of equation (8.11). Equation (11.3) is equivalent to

$$x_i = a'_{ij}(\bar{x}_j - b_j), \tag{11.4}$$

where a'_{ij} is the ij^{th} element of A'. But $a'_{ij} = a_{ji}$ and hence

$$x_i = a_{ji}(\bar{x}_j - b_j). \tag{11.5}$$

It now follows that

$$\frac{\partial x_i}{\partial \bar{x}_j} = a_{ji} \tag{11.6}$$

and hence that

$$\frac{\partial \bar{V}}{\partial \bar{x}_i} = \frac{\partial V}{\partial x_j}\frac{\partial x_j}{\partial \bar{x}_i} = a_{ij}\frac{\partial V}{\partial x_j}, \tag{11.7}$$

proving that $\partial V/\partial x_i$ is a vector. It is called the *gradient* of V and is denoted by grad V or ∇V.

If a tensor $A_{ij...}$ is defined at every point of some region of $\mathscr{E}_N$, the result is a *tensor field*. The partial derivatives $\partial A_{ij...}/\partial x_r$ can now be formed and constitute a tensor whose rank is one greater than that of $A_{ij...}$. We shall prove this for a second rank tensor field A_{ij}. The argument is easily made general. We have

$$\begin{aligned}\frac{\partial \bar{A}_{ij}}{\partial \bar{x}_k} &= \frac{\partial}{\partial \bar{x}_k}(a_{ir}a_{js}A_{rs}), \\ &= \frac{\partial}{\partial x_t}(a_{ir}a_{js}A_{rs})\frac{\partial x_t}{\partial \bar{x}_k}, \\ &= a_{ir}a_{js}a_{kt}\frac{\partial A_{rs}}{\partial x_t},\end{aligned} \tag{11.8}$$

by equation (11.6).

12. Contraction. Scalar product. Divergence

If two indices are made identical, a summation is implied. Thus, consider A_{ijk}. Then

$$A_{ijj} = A_{i11} + A_{i22} + \ldots + A_{iNN}. \tag{12.1}$$

ORTHOGONAL TRANSFORMATIONS 31

There are N^3 quantities A_{ijk}. However, of the indices in A_{ijj}, only i remains free to range over the integers $1, 2, \ldots, N$, and hence there are but N quantities A_{ijj} and we could put $B_i = A_{ijj}$. The rank has been reduced by two and the process is accordingly referred to as *contraction*.

Contraction of a tensor yields another tensor. For example, if $B_i = A_{ijj}$ then, employing equations (9.2),

$$\bar{B}_i = \bar{A}_{ijj} = a_{iq} a_{jr} a_{js} A_{qrs} = a_{iq} \delta_{rs} A_{qrs} = a_{iq} A_{qrr} = a_{iq} B_q. \quad (12.2)$$

Thus B_i is a vector. The argument is easily generalized.

In the special case of a tensor of rank two, e.g. A_{ij}, it follows that $\bar{A}_{ii} = A_{ii}$, i.e. A_{ii} is an invariant. Now, if A_i, B_i are vectors, $A_i B_j$ is a tensor. Hence, $A_i B_i$ is an invariant. This contracted product is called the *inner product* or the *scalar product* of the two vectors. We shall write

$$A_i B_i = \mathbf{A} \cdot \mathbf{B}. \quad (12.3)$$

In particular, the scalar product of a vector with itself is an invariant. The positive square root of this invariant will be called the *magnitude* of the vector. Thus, if A is the magnitude of A_i, then

$$A^2 = A_i A_i = \mathbf{A} \cdot \mathbf{A} = \mathbf{A}^2. \quad (12.4)$$

In $\mathscr{E}_3$, if θ is the angle between two vectors $\mathbf{A}$ and $\mathbf{B}$, then

$$AB \cos \theta = \mathbf{A} \cdot \mathbf{B}. \quad (12.5)$$

In $\mathscr{E}_N$, this equation is used to *define* θ. Hence, if

$$\mathbf{A} \cdot \mathbf{B} = 0, \quad (12.6)$$

then $\theta = \tfrac{1}{2}\pi$ and the vectors $\mathbf{A}$, $\mathbf{B}$ are said to be *orthogonal*.

If A_i is a vector field, $\partial A_i / \partial x_j$ is a tensor. By contraction it follows that $\partial A_i / \partial x_i$ is an invariant. This invariant is called the *divergence* of $\mathbf{A}$ and is denoted by div $\mathbf{A}$. Thus

$$\operatorname{div} \mathbf{A} = \frac{\partial A_i}{\partial x_i}. \quad (12.7)$$

More generally, if $A_{ij\ldots}$ is a tensor field, $\partial A_{ij\ldots} / \partial x_r$ is a tensor. This tensor derivative can now be contracted with respect to the index r

and any other index to yield another tensor, e.g. $\partial A_{ij\ldots}/\partial x_j$. This contraction is also referred to as the divergence of $A_{ij\ldots}$ with respect to the index j and we shall write

$$\frac{\partial A_{ij\ldots}}{\partial x_j} = \text{div}_j A_{ij\ldots}. \tag{12.8}$$

13. Tensor densities

$\mathfrak{A}_{ij}$ is a tensor density if, when the coordinates are subjected to the transformation (8.1), its components transform according to the law

$$\bar{\mathfrak{A}}_{ij} = |A| a_{ik} a_{jl} \mathfrak{A}_{kl}, \tag{13.1}$$

$|A|$ being the determinant of the transformation matrix A. Since for orthogonal transformations $|A| = \pm 1$ (equation (8.10)), relative to rectangular Cartesian frames, tensors and tensor densities are identical except that, for certain changes of axes, all the components of a density will be reversed in sign. For example, if in $\mathscr{E}_3$ a change is made from the right-handed system of axes to a left-handed system, the determinant of the transformation will be -1 and the components of a density will then be subject to this additional sign change.

Let $e_{ij\ldots n}$ be a density of the Nth rank which is skew-symmetric with respect to every pair of indices. Then all its components are zero, except those for which the indices $i, j, \ldots, n$ are all different and form a permutation of the numbers $1, 2, \ldots, N$. The effect of transposing any pair of indices in $e_{ij\ldots n}$ is to change its sign. It follows that if the arrangement $i, j, \ldots, n$ can be obtained from $1, 2, \ldots, N$ by an even number of transpositions, then $e_{ij\ldots n} = +e_{12\ldots N}$, whereas if it can be obtained by an odd number $e_{ij\ldots n} = -e_{12\ldots N}$. Relative to the x_i-axes let $e_{12\ldots N} = 1$. Then, in this frame, $e_{ij\ldots n}$ is 0 if $i, j, \ldots, n$ is not a permutation of $1, 2, \ldots, N$, is $+1$ if it is an even permutation and is -1 if it is an odd permutation. Transforming to the $\bar{x}_i$-axes, we find that

$$\bar{e}_{12\ldots N} = |A| a_{1i} a_{2j} \ldots a_{Nn} e_{ij\ldots n} = |A|^2 = 1. \tag{13.2}$$

But $\bar{e}_{ij\ldots n}$ is also skew-symmetric with respect to all its indices, since this is a property preserved by the transformation. Its components are also $0, \pm 1$ therefore and $e_{ij\ldots n}$ is a density with the same components in all frames. It is called the *Levi-Civita tensor density*.

ORTHOGONAL TRANSFORMATIONS

It may be shown without difficulty that:
 (i) the sum or difference of two densities of the same rank is a density,
 (ii) the product of a tensor and a density is a density,
 (iii) the product of a density and a density is a tensor,
 (iv) the partial derivative of a density with respect to x_i is a density,
 (v) a contracted density is a density.

Thus, to prove (iii), let $\mathfrak{A}_i$, $\mathfrak{B}_i$ be two vector densities. Then

$$\mathfrak{\bar{A}}_i \mathfrak{\bar{B}}_j = |A|^2 a_{ik} a_{jl} \mathfrak{A}_k \mathfrak{B}_l = a_{ik} a_{jl} \mathfrak{A}_k \mathfrak{B}_l. \tag{13.3}$$

The method is clearly quite general. The remaining results will be left as exercises for the reader.

14. Vector products. Curl

Throughout this section we shall be assuming that $N = 3$, i.e. the space will be ordinary Euclidean space.

Let A_{ij} be a skew-symmetric tensor. Defining

$$\mathfrak{A}_i = \tfrac{1}{2} e_{ijk} A_{jk}, \tag{14.1}$$

$\mathfrak{A}_i$ is a vector density. Its components are

$$\left.\begin{aligned}
\mathfrak{A}_1 &= \tfrac{1}{2}(A_{23} - A_{32}) = A_{23}, \\
\mathfrak{A}_2 &= \tfrac{1}{2}(A_{31} - A_{13}) = A_{31}, \\
\mathfrak{A}_3 &= \tfrac{1}{2}(A_{12} - A_{21}) = A_{12},
\end{aligned}\right\} \tag{14.2}$$

i.e. are the three distinct non-zero components of A_{ij}. We have proved, therefore, that these three components of any skew-symmetric tensor of the second rank may be regarded as the components of a vector density.

It is easy to verify that the inverse relationship to (14.1) is

$$A_{ij} = e_{ijk} \mathfrak{A}_k. \tag{14.3}$$

Let A_i, B_i be two vectors. From these we can form a skew-symmetric tensor of the second rank C_{ij} such that

$$C_{ij} = A_i B_j - A_j B_i. \tag{14.4}$$

From C_{ij} we can then form a vector density, viz.

$$\mathfrak{C}_i = \tfrac{1}{2} \mathrm{e}_{ijk} C_{jk} = \tfrac{1}{2} \mathrm{e}_{ijk}(A_j B_k - A_k B_j) = \mathrm{e}_{ijk} A_j B_k, \qquad (14.5)$$

whose components are

$$\left. \begin{array}{l} \mathfrak{C}_1 = A_2 B_3 - A_3 B_2, \\ \mathfrak{C}_2 = A_3 B_1 - A_1 B_3, \\ \mathfrak{C}_3 = A_1 B_2 - A_2 B_1. \end{array} \right\} \qquad (14.6)$$

Provided we employ only right-handed systems of axes or only left-handed systems in $\mathscr{E}_3$, $\mathfrak{C}_i$ is indistinguishable from a vector. If, however, a change is made from a left-handed system to a right-handed system, or vice versa, the components of $\mathfrak{C}_i$ are multiplied by -1 in addition to the usual vector transformation. Since it is usual to employ only right-handed frames, $\mathfrak{C}_i$ is often referred to as a vector (or an *axial vector*) and treated as such. It is then called the *vector product* of **A** and **B** and we write

$$\mathfrak{C} = \mathbf{A} \times \mathbf{B}. \qquad (14.7)$$

Vector multiplication is non-commutative, for

$$\mathbf{B} \times \mathbf{A} = \mathrm{e}_{ijk} B_j A_k = -\mathrm{e}_{ikj} A_k B_j = -\mathbf{A} \times \mathbf{B}, \qquad (14.8)$$

having made use of $\mathrm{e}_{ikj} = -\mathrm{e}_{ijk}$. However, vector multiplication obeys the distributive law, for

$$\mathbf{A} \times (\mathbf{B} + \mathbf{C}) = \mathrm{e}_{ijk} A_j (B_k + C_k) = \mathrm{e}_{ijk} A_j B_k + \mathrm{e}_{ijk} A_j C_k$$
$$= \mathbf{A} \times \mathbf{B} + \mathbf{A} \times \mathbf{C}. \qquad (14.9)$$

Again, if A_i is a vector field, we can construct the skew-symmetric tensor of the second rank

$$R_{ij} = \frac{\partial A_j}{\partial x_i} - \frac{\partial A_i}{\partial x_j} = A_{j,i} - A_{i,j}. \qquad (14.10)$$

We have introduced here the abbreviated notation $A_{i,j} = \partial A_i / \partial x_j$ for partial derivatives with respect to the coordinates. This will be made use of in many later arguments. Corresponding to R_{ij}, there is the vector density $\mathfrak{R}_i$ where

$$\mathfrak{R}_i = \tfrac{1}{2} \mathrm{e}_{ijk} R_{jk} = \tfrac{1}{2} \mathrm{e}_{ijk}(A_{k,j} - A_{j,k}) = \mathrm{e}_{ijk} A_{k,j}. \qquad (14.11)$$

ORTHOGONAL TRANSFORMATIONS

This has components

$$\left.\begin{aligned}\mathfrak{R}_1 &= \frac{\partial A_3}{\partial x_2} - \frac{\partial A_2}{\partial x_3}, \\ \mathfrak{R}_2 &= \frac{\partial A_1}{\partial x_3} - \frac{\partial A_3}{\partial x_1}, \\ \mathfrak{R}_3 &= \frac{\partial A_2}{\partial x_1} - \frac{\partial A_1}{\partial x_2},\end{aligned}\right\} \quad (14.12)$$

and is denoted by curl **A**. It, also, is an axial vector.

Equation (14.5) can still be employed to define a vector product when either or both of the vectors **A**, **B** are replaced by vector densities. If only one is replaced by a vector density, the right-hand member of equation (14.5) will involve the product of two densities and a vector. The resulting vector product will then be a vector. Similarly, by replacing **A** in equation (14.11) by a vector density, the curl of a vector density is defined as an ordinary vector.

Exercises 2

1. Show that, in two dimensions, the general orthogonal transformation has matrix A given by

$$A = \begin{pmatrix} \cos\theta & \sin\theta \\ -\sin\theta & \cos\theta \end{pmatrix}.$$

Verify that $|A| = 1$ and that $A^{-1} = A'$. T_{ij} is a tensor in this space. Write down in full the transformation equations for all its components and deduce that T_{ii} is an invariant.

2. $\bar{x} = Ax$, $\bar{\bar{x}} = B\bar{x}$ are two successive orthogonal transformations relative to each of which T_{ij} transforms as a tensor. Show that the resultant transformation $\bar{\bar{x}} = BAx$ is orthogonal and that T_{ij} transforms as a tensor with respect to it.

3. Show that contraction of the Levi-Civita density results in the zero tensor density.

4. In $\mathscr{E}_3$, prove that

$$\operatorname{curl} \operatorname{grad} V = 0, \quad \operatorname{div} \operatorname{curl} \mathbf{A} = 0.$$

5. In $\mathscr{E}_3$, prove that

(i) $\qquad e_{ikl}\,e_{imn} = \delta_{km}\delta_{ln} - \delta_{kn}\delta_{lm}$,

(ii) $\qquad e_{ikl}\,e_{ikm} = 2\delta_{lm}$.

6. In $\mathscr{E}_3$, show that

$$\nabla^2 V = \operatorname{div}\operatorname{grad} V = \frac{\partial^2 V}{\partial x_1^2} + \frac{\partial^2 V}{\partial x_2^2} + \frac{\partial^2 V}{\partial x_3^2} = \frac{\partial^2 V}{\partial x_i\,\partial x_i}.$$

7. In $\mathscr{E}_3$, prove that

$$\operatorname{curl}\operatorname{curl} \mathbf{A} = \operatorname{grad}\operatorname{div} \mathbf{A} - \nabla^2 \mathbf{A}.$$

[Hint: Employ Exercise 5 (i).]

8. In $\mathscr{E}_3$, prove that

(i) $\qquad \mathbf{A}\times(\mathbf{B}\times\mathbf{C}) = \mathbf{A}\cdot\mathbf{C}\,\mathbf{B} - \mathbf{A}\cdot\mathbf{B}\,\mathbf{C},$

(ii) $\qquad \mathbf{A}\cdot\mathbf{B}\times\mathbf{C} = \begin{vmatrix} A_1 & B_1 & C_1 \\ A_2 & B_2 & C_2 \\ A_3 & B_3 & C_3 \end{vmatrix}.$

9. In $\mathscr{E}_N$, prove that

$$\operatorname{div} V\mathbf{A} = V\operatorname{div}\mathbf{A} + \mathbf{A}\cdot\operatorname{grad} V.$$

10. In $\mathscr{E}_3$, prove that

(i) $\quad \operatorname{curl} V\mathbf{A} = V\operatorname{curl}\mathbf{A} - \mathbf{A}\times\operatorname{grad} V,$

(ii) $\quad \operatorname{div}(\mathbf{A}\times\mathbf{B}) = \mathbf{B}\cdot\operatorname{curl}\mathbf{A} - \mathbf{A}\cdot\operatorname{curl}\mathbf{B},$

(iii) $\quad \operatorname{curl}(\mathbf{A}\times\mathbf{B}) = \mathbf{B}\cdot\nabla\mathbf{A} - \mathbf{A}\cdot\nabla\mathbf{B} + \mathbf{A}\operatorname{div}\mathbf{B} - \mathbf{B}\operatorname{div}\mathbf{A},$

(iv) $\quad \operatorname{grad}(\mathbf{A}\cdot\mathbf{B}) = \mathbf{B}\cdot\nabla\mathbf{A} + \mathbf{A}\cdot\nabla\mathbf{B} + \mathbf{A}\times\operatorname{curl}\mathbf{B} + \mathbf{B}\times\operatorname{curl}\mathbf{A},$

where $\qquad \mathbf{A}\cdot\nabla\mathbf{B} = A_j B_{i,j}.$

11. If A_{ij} is a tensor and $B_{ij} = A_{ji}$, prove that B_{ij} is a tensor. Deduce that if A_{ij} is symmetric in one frame, it is so in all.

12. Prove that $\delta_{ij}\delta_{ik} = \delta_{jk}$

and that $e_{ijk}e_{lmn}$ has the value $+1$ if i, j, k are all different and (lmn) is an even permutation of (ijk), -1 if i, j, k are all different and (lmn) is an odd permutation of (ijk), and 0 otherwise. Deduce that

$$e_{ijk}e_{lmn} = \delta_{il}\delta_{jm}\delta_{kn} + \delta_{im}\delta_{jn}\delta_{kl} + \delta_{in}\delta_{jl}\delta_{km} \\ - \delta_{in}\delta_{jm}\delta_{kl} - \delta_{il}\delta_{jn}\delta_{km} - \delta_{im}\delta_{jl}\delta_{kn}.$$

Hence prove that

$$e_{ijk}e_{imn} = \delta_{jm}\delta_{kn} - \delta_{jn}\delta_{km}.$$

(M. T.)

13. In $\mathscr{E}_3$, prove that

(i) $(\mathbf{a}\times\mathbf{b})\cdot(\mathbf{c}\times\mathbf{d}) = \mathbf{a}\cdot\mathbf{c}\,\mathbf{b}\cdot\mathbf{d} - \mathbf{a}\cdot\mathbf{d}\,\mathbf{b}\cdot\mathbf{c},$

(ii) $(\mathbf{a}\times\mathbf{b})\times(\mathbf{c}\times\mathbf{d}) = [\mathbf{acd}]\mathbf{b} - [\mathbf{bcd}]\mathbf{a}$
$= [\mathbf{abd}]\mathbf{c} - [\mathbf{abc}]\mathbf{d},$

where $[\mathbf{abc}] = \mathbf{a}\cdot\mathbf{b}\times\mathbf{c}$.

CHAPTER 3

Special Relativity Mechanics

15. The velocity vector

Suppose that a point P is in motion relative to an inertial frame S. Let ds be the distance between successive positions of P which it occupies at times t, $t+dt$ respectively. Then, by equation (7.4), if $d\tau$ is the proper time interval between these two events,

$$d\tau = \left(dt^2 - \frac{1}{c^2} ds^2\right)^{1/2} = (1-v^2/c^2)^{1/2} dt, \qquad (15.1)$$

where $v = ds/dt$ is the speed of P as measured in S. Now, as shown in section 7, $d\tau$ is the time interval between the two events as measured in a frame for which the events occur at the same point. Thus $d\tau$ is the time interval measured by a clock moving with P. dt is the time interval measured by clocks stationary in S. Equation (15.1) indicates that, as observed from S, the rate of the clock moving with P is slow by a factor $(1-v^2/c^2)^{1/2}$. This is the phenomenon of time dilation already commented upon in section 6. If P leaves a point A at $t=t_1$ and arrives at a point B at $t=t_2$, the time of transit as registered by a clock moving with P will be

$$\tau_2 - \tau_1 = \int_{t_1}^{t_2} (1-v^2/c^2)^{1/2} dt. \qquad (15.2)$$

The successive positions of P together with the times it occupies these positions constitute a series of events which will lie on the point's world-line in Minkowski space-time. Erecting rectangular axes in space-time corresponding to the rectangular Cartesian frame S, let x_i, $x_i + dx_i$ be the coordinates of adjacent points on the world-line. These points will represent the events (x,y,z,t), $(x+dx, y+dy, z+dz, t+dt)$ in S. If (v_x, v_y, v_z) are the components of the velocity vector $\mathbf{v}$ of P relative to S, then

$$v_x = \frac{dx}{dt}, \quad v_y = \frac{dy}{dt}, \quad v_z = \frac{dz}{dt}. \qquad (15.3)$$

SPECIAL RELATIVITY MECHANICS 39

v does not possess the transformation properties of a vector relative to orthogonal transformations (i.e. Lorentz transformations) in space-time. It is a vector relative to rectangular axes stationary in S only. However, we can define a *4-velocity vector* which does possess such properties as follows: dx_i is a displacement vector relative to rectangular axes in space-time and $d\tau$ is an invariant. It follows that $dx_i/d\tau$ is a vector relative to Lorentz transformations expressed as orthogonal transformations in space-time. It is called the 4-velocity vector of P and will be denoted by **V**.

V can be expressed in terms of **v** thus:

$$\frac{dx_i}{d\tau} = \frac{dx_i}{dt}\frac{dt}{d\tau} = (1-v^2/c^2)^{-1/2}\dot{x}_i \tag{15.4}$$

by equation (15.1). Also, from equations (4.4) we obtain

$$\dot{x}_1 = v_x, \quad \dot{x}_2 = v_y, \quad \dot{x}_3 = v_z, \quad \dot{x}_4 = ic. \tag{15.5}$$

It now follows from these equations that

$$\mathbf{V} = (1-v^2/c^2)^{-1/2}(v_x, v_y, v_z, ic) = (1-v^2/c^2)^{-1/2}(\mathbf{v}, ic), \tag{15.6}$$

where the notation should be clear without further explanation.

Knowing the manner in which the components of **V** transform when new axes are chosen in space-time, equation (15.6) enables us to calculate how the components of **v** transform when S is replaced by a new inertial frame $\bar{S}$. Thus, consider the orthogonal transformation (5.1) which has been interpreted as a change from an inertial frame S to another $\bar{S}$ related to the first as shown in Fig. 2. The corresponding transformation equations for **V** are

$$\left.\begin{aligned}\bar{V}_1 &= V_1\cos\alpha + V_4\sin\alpha, & \bar{V}_2 &= V_2, \\ \bar{V}_4 &= -V_1\sin\alpha + V_4\cos\alpha, & \bar{V}_3 &= V_3.\end{aligned}\right\} \tag{15.7}$$

By equation (15.6), these equations are equivalent to

$$\left.\begin{aligned}(1-\bar{v}^2/c^2)^{-1/2}\bar{v}_x &= (1-v^2/c^2)^{-1/2}(v_x\cos\alpha + ic\sin\alpha),\\ (1-\bar{v}^2/c^2)^{-1/2}\bar{v}_y &= (1-v^2/c^2)^{-1/2}v_y,\\ (1-\bar{v}^2/c^2)^{-1/2}\bar{v}_z &= (1-v^2/c^2)^{-1/2}v_z,\\ (1-\bar{v}^2/c^2)^{-1/2}ic &= (1-v^2/c^2)^{-1/2}(-v_x\sin\alpha + ic\cos\alpha),\end{aligned}\right\} \tag{15.8}$$

where $\bar{\mathbf{v}}$ is the velocity of the point as measured in the frame $\bar{S}$. Substituting for $\cos\alpha$, $\sin\alpha$ from equations (5.7), equations (15.8) can be written

$$\left. \begin{array}{l} \bar{v}_x = Q(v_x - u), \\ \bar{v}_y = Q(1 - u^2/c^2)^{1/2} v_y, \\ \bar{v}_z = Q(1 - u^2/c^2)^{1/2} v_z, \\ 1 = Q(1 - uv_x/c^2), \end{array} \right\} \quad (15.9)$$

where
$$Q = \left[\frac{1 - \bar{v}^2/c^2}{(1 - v^2/c^2)(1 - u^2/c^2)} \right]^{1/2} \quad (15.10)$$

Dividing the first three equations (15.9) by the fourth, we obtain the special Lorentz transformation equations for **v** in their final form, viz.

$$\left. \begin{array}{l} \bar{v}_x = \dfrac{v_x - u}{1 - uv_x/c^2}, \\[4pt] \bar{v}_y = \dfrac{(1 - u^2/c^2)^{1/2} v_y}{1 - uv_x/c^2}, \\[4pt] \bar{v}_z = \dfrac{(1 - u^2/c^2)^{1/2} v_z}{1 - uv_x/c^2}. \end{array} \right\} \quad (15.11)$$

If u and v are small by comparison with c, equations (15.11) can be replaced by the approximate equations

$$\bar{v}_x = v_x - u, \quad \bar{v}_y = v_y, \quad \bar{v}_z = v_z. \quad (15.12)$$

These are equivalent to the vector equation (1.1) relating velocity measurements in two inertial frames according to Newtonian mechanics.

Since, by the fourth of equations (15.9) Q must be real, equation (15.10) implies that if $\bar{v} < c$ then $v < c$. Thus, if a point is moving with a velocity approaching c in $\bar{S}$ and $\bar{S}$ is moving relative to S with a velocity of the same order, the point's velocity relative to S will still be less than c. Such a result is, of course, completely at variance with classical ideas. In particular, if a light pulse is being propagated along Ox so that $v_x = c$, $v_y = v_z = 0$, then it will be found that $\bar{v}_x = c$,

SPECIAL RELATIVITY MECHANICS

$\bar{v}_y = \bar{v}_z = 0$. This confirms that light is propagated with speed c in all inertial frames.

The transformation inverse to (15.11) can be found by exchanging 'barred' and 'unbarred' velocity components and replacing u by $-u$.

16. Mass and momentum

In section 2 it was shown that Newton's laws of motion conform to the special principle of relativity. However, the argument involved classical ideas concerning space-time relationships between two inertial frames and these have since been replaced by relationships based upon the Lorentz transformation. The whole question must therefore be re-examined and this we shall do in this and the following section.

We shall begin by considering the conservation of momentum equation (1.3) for the impact of two particles by which mass is defined in classical mechanics. Since the velocity vectors $\mathbf{u}_1$, etc. are not vectors relative to orthogonal transformations in space-time and indeed transform between inertial frames in a very complex manner, it is at once evident that equation (1.3) is not covariant with respect to transformations between inertial frames. It will accordingly be replaced, tentatively, by another equation, viz.

$$M_1 \mathbf{U}_1 + M_2 \mathbf{U}_2 = M_1 \mathbf{V}_1 + M_2 \mathbf{V}_2, \qquad (16.1)$$

where $\mathbf{U}_1$, etc. are the 4-velocities of the particles and M_1, M_2 are invariants associated with the particles which will correspond to their classical masses. This is a vector equation and hence is covariant with respect to orthogonal transformations in space-time as we require. Equation (16.1) will be abbreviated to the statement

$$\sum M\mathbf{V} \text{ is conserved} \qquad (16.2)$$

and then, by equation (15.6), this implies that

$$\sum m(\mathbf{v}, ic) \text{ is conserved}, \qquad (16.3)$$

where
$$m = \frac{M}{(1-v^2/c^2)^{1/2}}. \qquad (16.4)$$

By consideration of the first three (or space) components of (16.3), it will be clear that

$$\sum m\mathbf{v} \text{ is conserved,} \tag{16.5}$$

and, by consideration of the fourth (or time) component that

$$\sum m \text{ is conserved.} \tag{16.6}$$

If, therefore, m is identified as the quantity which will play the role of the Newtonian mass in special relativity mechanics, our tentative conservation law (16.1) is seen to incorporate both the principles of conservation of momentum and of mass from Newtonian mechanics. The principle (16.1) is accordingly eminently reasonable. However, our ultimate justification for accepting it is, of course, that its consequences are verified experimentally. We shall refer to such checks at appropriate points in the later development.

It appears from equation (16.4) that the mass of a particle must now be regarded as being dependent upon its speed v. If $v = 0$, then $m = M$. Thus M is the mass of the particle when measured in an inertial frame in which it is stationary. M will be referred to as the *rest mass* or *proper mass* and will, in future, be denoted by m_0. Then

$$m = \frac{m_0}{(1-v^2/c^2)^{1/2}}. \tag{16.7}$$

Clearly $m \to \infty$ as $v \to c$, implying that inertia effects become increasingly serious as the velocity of light is approached and prevent this velocity being attained by any material particle. This is in agreement with our earlier observations. Formula (16.7) has been verified by observation of collisions between atomic nuclei and cosmic ray particles (e.g. see Exercise 14 at the end of this chapter).

We shall define the *4-momentum vector* $\mathbf{P}$ of a particle whose proper mass is m_0 and whose 4-velocity is $\mathbf{V}$, by the equation

$$\mathbf{P} = m_0 \mathbf{V}. \tag{16.8}$$

Since m_0 is an invariant and $\mathbf{V}$ is a vector in space-time, $\mathbf{P}$ is a vector. By equation (15.6),

$$\mathbf{P} = m_0(1-v^2/c^2)^{-1/2}(\mathbf{v}, ic) = (m\mathbf{v}, imc) = (\mathbf{p}, imc), \tag{16.9}$$

where $\mathbf{p} = m\mathbf{v}$ is the classical momentum.

SPECIAL RELATIVITY MECHANICS   43

Relative to the special orthogonal transformation (5.1), the transformation equations for the components of **P** are

$$\left.\begin{array}{ll} \bar{P}_1 = P_1\cos\alpha + P_4\sin\alpha, & \bar{P}_2 = P_2, \\ \bar{P}_4 = -P_1\sin\alpha + P_4\cos\alpha, & \bar{P}_3 = P_3. \end{array}\right\} \quad (16.10)$$

Substituting for the components of **P** from equation (16.9) and similarly for **P̄**, and employing equations (5.7), it will be found that

$$\left.\begin{array}{l} \bar{p}_x = \dfrac{p_x - mu}{(1-u^2/c^2)^{1/2}}, \\ \bar{p}_y = p_y, \\ \bar{p}_z = p_z, \end{array}\right\} \quad (16.11)$$

$$\bar{m} = \frac{m - p_x u/c^2}{(1-u^2/c^2)^{1/2}}. \quad (16.12)$$

Equations (16.11) constitute the special Lorentz transformation equations for the components of the momentum **p** and equation (16.12) the corresponding transformation equation for mass. Since $p_x = mv_x$, this equation can also be written

$$\bar{m} = \frac{1 - uv_x/c^2}{(1-u^2/c^2)^{1/2}} m. \quad (16.13)$$

This reduces to the classical form of equation (2.4) if u, v_x are negligible by comparison with c.

17. The force vector. Energy

We have seen that in classical mechanics, when the mass of a particle has been determined, the force acting upon it at any instant is specified by Newton's second law. Force receives a similar definition in special relativity mechanics. The mass of a particle with a given velocity can be determined by permitting it to collide with a standard particle and applying the principle of momentum conservation. Equation (16.7) then gives its mass at any velocity. The force **f** acting upon a particle having mass m and velocity **v** relative to some inertial frame is then defined by the equation,

$$\mathbf{f} = \frac{d}{dt}(m\mathbf{v}) = \frac{d\mathbf{p}}{dt}, \quad (17.1)$$

where **p** is the particle's momentum. Clearly **f** will be dependent upon the inertial frame employed, a departure from classical mechanics.

Definition (17.1) implies that, if equal and opposite forces act upon two colliding particles, momentum is conserved. However, although experiment confirms that momentum is indeed conserved, Newton's third law cannot be incorporated in the new mechanics, for it will appear later that, if the forces are equal and opposite for one inertial observer, in general they are not so for all such observers. Equation (16.1) therefore replaces this law in the new mechanics.

f is not a vector with respect to Lorentz transformations in space-time. However, a *4-force* **F** can be defined which has this property. The natural definition is clearly

$$\mathbf{F} = \frac{d\mathbf{P}}{d\tau} = m_0 \frac{d\mathbf{V}}{d\tau}, \qquad (17.2)$$

P being the 4-momentum and τ the proper time for the particle. **F** is immediately expressible in terms of **f** for, by equation (16.9),

$$\begin{aligned} \mathbf{F} &= \frac{d}{d\tau}(\mathbf{p}, imc), \\ &= \frac{d}{dt}(\mathbf{p}, imc)\frac{dt}{d\tau}, \\ &= (1 - v^2/c^2)^{-1/2}(\dot{\mathbf{p}}, i\dot{m}c), \\ &= (1 - v^2/c^2)^{-1/2}(\mathbf{f}, i\dot{m}c). \end{aligned} \qquad (17.3)$$

The vectors **V**, **F** are orthogonal. This is proved as follows: From equation (15.6)

$$\mathbf{V}^2 = -c^2. \qquad (17.4)$$

Differentiating with respect to τ,

$$\mathbf{V} \cdot \frac{d\mathbf{V}}{d\tau} = 0,$$

i.e. $$\mathbf{V} \cdot \mathbf{F} = 0, \qquad (17.5)$$

as stated. This result has very important consequences. Substituting for **V** and **F** from equations (15.6) and (17.3) respectively, it is clear that

$$(1-v^2/c^2)^{-1}(\mathbf{v}, ic)\cdot(\mathbf{f}, i\dot{m}c) = 0. \tag{17.6}$$

This is equivalent to
$$\mathbf{v}\cdot\mathbf{f} - c^2 \dot{m} = 0. \tag{17.7}$$

But, by definition, $\mathbf{v}\cdot\mathbf{f}$ is the rate at which $\mathbf{f}$ is doing work. It follows that the work done by the force acting on the particle during a time interval (t_1, t_2) is

$$\int_{t_1}^{t_2} c^2 \dot{m}\, dt = m_2 c^2 - m_1 c^2. \tag{17.8}$$

The classical equation of work is

$$\text{work done} = \text{increase in kinetic energy}, \tag{17.9}$$

where $T = \tfrac{1}{2} m v^2$ is the kinetic energy. Equation (17.8) indicates that in special relativity mechanics we must define T by a formula of the type

$$T = mc^2 + \text{constant}. \tag{17.10}$$

When $v = 0$, $T = 0$ and this determines the unknown constant to be $-m_0 c^2$. Thus

$$T = \frac{m_0 c^2}{(1-v^2/c^2)^{1/2}} - m_0 c^2. \tag{17.11}$$

If v/c is small, $(1-v^2/c^2)^{-1/2} = 1 + v^2/2c^2$ approximately and the above equation reduces to $T = \tfrac{1}{2} m_0 v^2$, in agreement with classical theory.

According to equation (17.10), any increase in the kinetic energy of a particle will result in a proportional increase in its mass. Thus, if a body is heated so that the thermal agitation of its molecules is increased, the masses of these particles, and hence the total body mass, will increase in proportion to the heat energy which has been communicated.

Again, suppose two equal elastic particles approach one another along the same straight line with equal speeds v. If their proper masses are both m_0, the net mass in the system before collision is

$$2m_0/(1-v^2/c^2)^{1/2}.$$

It has been accepted as a fundamental principle that this mass will be conserved during the collision. However, from considerations of symmetry, it is obvious that at some instant during the impact both particles will be brought to rest and their masses at this instant will be proper masses m_0'. By our principle,

$$2m_0' = \frac{2m_0}{(1-v^2/c^2)^{1/2}}. \qquad (17.12)$$

It follows, therefore, that, at this instant, the proper mass of each particle has increased by

$$\frac{m_0}{(1-v^2/c^2)^{1/2}} - m_0 = T/c^2, \qquad (17.13)$$

where T is the original KE of the particle and use has been made of equation (17.11). Now, in losing this KE, the particle has had an equal amount of work done upon it by the force of interaction and this has resulted in a distortion of the elastic material of which it is made. At the instant each particle is brought to rest, this distortion is at a maximum and the elastic potential energy as measured by the work done will be exactly T. If we assume that this increase in the internal energy of the particle leads to a proportional increase in mass, the increment of rest mass (17.13) is explained. If the particles are not perfectly elastic, the work done in bringing them to rest will not only increase the internal elastic energy, but will also generate heat. Both forms of energy will then contribute to increase the proper masses.

Such considerations as these suggest very strongly that mass and energy are equivalent, being two different measures of the same physical quantity. Thus, the distinction between mass and energy which was maintained in classical physical theories, has now been abandoned. All forms of energy E, mechanical, thermal, electromagnetic, are now taken to possess inertia of mass m, according to *Einstein's Equation*, viz.

$$E = mc^2. \qquad (17.14)$$

SPECIAL RELATIVITY MECHANICS 47

Conversely, any particle whose mass is m, has associated energy E and, by equation (17.11),

$$E = T + m_0 c^2. \tag{17.15}$$

$m_0 c^2$ is interpreted as the internal energy of the particle when stationary. If the particle were converted completely into electro-magnetic radiation, $m_0 c^2$ would be the energy released. This is the source of the energy released in an atomic explosion. The mass of the material products of the explosion is slightly less than the net mass present before the explosion, the difference being accounted for by the mass of the energy released. Even a small mass deficiency implies that an immense quantity of energy has been released. Thus, if $m = 1$ gm, $c = 3 \times 10^{10}$ cm/sec and hence $E = 9 \times 10^{20}$ ergs $= 2 \cdot 5 \times 10^7$ kilowatt hours.

The principle of conservation of mass, which has been incorporated into the new mechanics, is now seen to be identical with the principle of conservation of energy, which is accordingly also regarded as valid in the new mechanics. However, the distinction between the two principles, which was a feature of the older mechanics, has disappeared.

18. Lorentz transformation equations for force
By equation (17.7),

$$i\dot{m}c = \frac{i}{c}\mathbf{f}\cdot\mathbf{v}. \tag{18.1}$$

Referring to equation (17.3), $\mathbf{F}$ can now be completely expressed in terms of $\mathbf{f}$ thus:

$$\mathbf{F} = (1 - v^2/c^2)^{-1/2}\left(\mathbf{f}, \frac{i}{c}\mathbf{f}\cdot\mathbf{v}\right). \tag{18.2}$$

Relative to the special Lorentz transformation, the transformation equations for the components of $\mathbf{F}$ are

$$\left.\begin{array}{ll} \bar{F}_1 = F_1 \cos\alpha + F_4 \sin\alpha, & \bar{F}_2 = F_2, \\ \bar{F}_4 = -F_1 \sin\alpha + F_4 \cos\alpha, & \bar{F}_3 = F_3. \end{array}\right\} \tag{18.3}$$

Substituting from equation (18.2) into the first three of these equations and employing equations (5.7), it follows that

$$\left.\begin{aligned} \bar{f}_x &= Q\left(f_x - \frac{u}{c^2}\mathbf{f}\cdot\mathbf{v}\right), \\ \bar{f}_y &= Q(1-u^2/c^2)^{1/2} f_y, \\ \bar{f}_z &= Q(1-u^2/c^2)^{1/2} f_z, \end{aligned}\right\} \quad (18.4)$$

where Q is given by equation (15.10). Substituting for Q from the fourth of equations (15.9), it will be found that

$$\left.\begin{aligned} \bar{f}_x &= f_x - \frac{u}{c^2}\cdot\frac{(f_y v_y + f_z v_z)}{1 - u v_x/c^2}, \\ \bar{f}_y &= \frac{(1-u^2/c^2)^{1/2}}{1 - u v_x/c^2} f_y, \\ \bar{f}_z &= \frac{(1-u^2/c^2)^{1/2}}{1 - u v_x/c^2} f_z. \end{aligned}\right\} \quad (18.5)$$

These are the special Lorentz transformation equations for $\mathbf{f}$. If u, v are negligible by comparison with c, these equations reduce to the classical form of equation (2.6).

It is clear from equations (18.5) that, if equal and opposite forces are observed from S to act upon two particles, the forces observed from $\bar{S}$ will not be so related unless the particles' velocities are the same.

19. Motion with variable proper mass

In section 17 it has been assumed that the proper mass m_0 of the particle which is moving under the action of the force $\mathbf{f}$, is constant throughout the motion. If, however, the particle is being heated or cooled during its motion, or if any non-mechanical forms of energy are being communicated to it from an external source, its proper mass will vary and our equations must be modified to take account of this variation.

Thus, consider equation (17.2). The 4-momentum $\mathbf{P}$ is still defined

SPECIAL RELATIVITY MECHANICS

by equation (16.8) but, since m_0 is variable, the 4-force is given by

$$\mathbf{F} = \frac{d}{d\tau}(m_0 \mathbf{V}) = m_0 \frac{d\mathbf{V}}{d\tau} + \mathbf{V}\frac{dm_0}{d\tau}. \tag{19.1}$$

Equation (17.4) remains valid and hence, differentiating,

$$\mathbf{V} \cdot \frac{d\mathbf{V}}{d\tau} = 0. \tag{19.2}$$

Substituting for $d\mathbf{V}/d\tau$ from equation (19.1), we obtain

$$\mathbf{V} \cdot \mathbf{F} = -c^2 \frac{dm_0}{d\tau}. \tag{19.3}$$

We conclude that $\mathbf{V}$ and $\mathbf{F}$ are no longer orthogonal vectors. Substituting in equation (19.3) for $\mathbf{V}$ and $\mathbf{F}$ from equations (15.6) and (17.3) respectively, it will be found that

$$\frac{dE}{dt} = \mathbf{f} \cdot \mathbf{v} + (c^2 - v^2)\frac{dm_0}{d\tau}. \tag{19.4}$$

This is the modified equation of work. Its physical interpretation is clearly:

rate of increase in particle's energy

= rate of doing work by applied force

+ rate of energy input from the external source. (19.5)

We deduce that energy is being taken from the external source at a rate

$$R = (c^2 - v^2)\frac{dm_0}{d\tau} \tag{19.6}$$

as measured in the inertial frame being employed.

Equation (19.4) can therefore be written

$$\dot{E} = c^2 \dot{m} = \mathbf{f} \cdot \mathbf{v} + R \tag{19.7}$$

and hence, by equation (17.3),

$$\mathbf{F} = (1 - v^2/c^2)^{-1/2} \left[\mathbf{f}, \frac{i}{c}(\mathbf{f} \cdot \mathbf{v} + R) \right]. \tag{19.8}$$

This is the modified form of equation (18.2).

20. Lagrange's and Hamilton's equations

Suppose that a particle having constant proper mass m_0 is in motion relative to an inertial frame under the action of a force derivable from a potential V. Then its equations of motion are

$$\frac{d}{dt}\left\{\frac{m_0 \dot{x}}{(1-v^2/c^2)^{1/2}}\right\} = -\frac{\partial V}{\partial x}, \text{ etc.} \tag{20.1}$$

Expressed in Lagrange form, these equations must be

$$\frac{d}{dt}\left(\frac{\partial L}{\partial \dot{x}}\right) = \frac{\partial L}{\partial x}, \text{ etc.} \tag{20.2}$$

and hence L must be a function of $x, y, z, \dot{x}, \dot{y}, \dot{z}$, such that

$$\frac{\partial L}{\partial \dot{x}} = \frac{m_0 \dot{x}}{(1-v^2/c^2)^{1/2}}, \quad \frac{\partial L}{\partial x} = -\frac{\partial V}{\partial x}, \text{ etc.} \tag{20.3}$$

Since $v^2 = \dot{x}^2 + \dot{y}^2 + \dot{z}^2$, these equations can be validated by taking

$$L = -m_0 c^2 (1-v^2/c^2)^{1/2} - V, \tag{20.4}$$

which is accordingly the Lagrangian for the particle.

Now

$$\frac{\partial L}{\partial \dot{x}} = p_x, \text{ etc.} \tag{20.5}$$

and it follows exactly as in classical theory that, if the Hamiltonian H is defined by the equation

$$H = p_x v_x + p_y v_y + p_z v_z - L, \tag{20.6}$$

and is then expressed as a function of the quantities x, y, z, p_x, p_y, p_z alone, the Lagrange equations (20.2) are equivalent to Hamilton's equations

$$\dot{x} = \frac{\partial H}{\partial p_x}, \quad \dot{p}_x = -\frac{\partial H}{\partial x}, \text{ etc.} \tag{20.7}$$

Now

$$p_x v_x + p_y v_y + p_z v_z = \frac{m_0 v^2}{(1-v^2/c^2)^{1/2}} \tag{20.8}$$

SPECIAL RELATIVITY MECHANICS

and hence

$$\begin{aligned} H &= \frac{m_0 v^2}{(1-v^2/c^2)^{1/2}} + m_0 c^2 (1-v^2/c^2)^{1/2} + V, \\ &= \frac{m_0 c^2}{(1-v^2/c^2)^{1/2}} + V, \\ &= E + V \end{aligned} \qquad (20.9)$$

the total energy, precisely as for classical theory.

But

$$\begin{aligned} p_x^2 + p_y^2 + p_z^2 &= \frac{m_0^2 v^2}{1-v^2/c^2}, \\ &= -m_0^2 c^2 + \frac{m_0^2 c^2}{1-v^2/c^2}, \\ &= -m_0^2 c^2 + E^2/c^2, \end{aligned} \qquad (20.10)$$

and it follows that

$$E^2 = c^2(p_x^2 + p_y^2 + p_z^2 + m_0^2 c^2). \qquad (20.11)$$

Substituting in equation (20.9)

$$H = c(p_x^2 + p_y^2 + p_z^2 + m_0^2 c^2)^{1/2} + V, \qquad (20.12)$$

expressing H as a function of x, y, z, p_x, p_y, p_z. The reader is now left to verify that Hamilton's equations are equivalent to the equations of motion (20.1).

Exercises 3

1. Obtain the transformation equations for **v** by differentiating the Lorentz transformation.

2. Obtain the transformation equations for the acceleration **a** by differentiating the transformation equations for **v** and express them in the form

$$\bar{a}_x = \frac{(1-u^2/c^2)^{3/2}}{(1-v_x u/c^2)^3} a_x,$$

$$\bar{a}_y = \frac{1-u^2/c^2}{(1-v_x u/c^2)^2}\left(a_y + \frac{v_y u/c^2}{1-v_x u/c^2} a_x\right),$$

$$\bar{a}_z = \frac{1-u/^2c^2}{(1-v_x u/c^2)^2}\left(a_z + \frac{v_z u/c^2}{1-v_x u/c^2} a_x\right).$$

Deduce that a point which has uniform acceleration in one inertial frame has not, in general, uniform acceleration in another.

3. If $\bar{S}$ has velocity c relative to S, show that all points moving relative to S with velocities less than c have a velocity c as observed from $\bar{S}$.

4. Two points are moving in opposite directions with speeds c relative to some inertial frame. Show that their relative velocity is c.

5. Show that the 4-velocity **V** is of constant magnitude ic.

6. A beam of light is being propagated in the xy-plane of S at an angle α to the x-axis. Relative to $\bar{S}$ it is observed to make an angle $\bar{\alpha}$ with $\bar{O}\bar{x}$. Prove the aberration of light formula, viz.

$$\cot \bar{\alpha} = \frac{\cot \alpha - (u/c) \csc \alpha}{(1-u^2/c^2)^{1/2}}.$$

Deduce that, if $u \ll c$, then

$$\Delta \alpha = \bar{\alpha} - \alpha = \frac{u}{c} \sin \alpha,$$

approximately.

7. A particle of proper mass m_0 is moving under the action of a force **f** with velocity **v**. Show that

$$\mathbf{f} = \frac{m_0}{(1-v^2/c^2)^{1/2}} \frac{d\mathbf{v}}{dt} + \frac{m_0 v\dot{v}/c^2}{(1-v^2/c^2)^{3/2}} \mathbf{v}.$$

Hence, if the acceleration $d\mathbf{v}/dt$ is parallel to **v**, show that

$$\mathbf{f} = \frac{m_0}{(1-v^2/c^2)^{3/2}} \frac{d\mathbf{v}}{dt},$$

and if the acceleration is perpendicular to **v**, then

$$\mathbf{f} = \frac{m_0}{(1-v^2/c^2)^{1/2}} \frac{d\mathbf{v}}{dt}.$$

8. Show that $\mathbf{P} = (\mathbf{p}, iE/c)$ and deduce that

$$p^2 - E^2/c^2$$

is an invariant $-m_0^2 c^2$ with respect to Lorentz transformations.

9. Show that E transforms under a special Lorentz transformation according to the equation

$$\bar{E} = \frac{E - p_x u}{(1 - u^2/c^2)^{1/2}}.$$

10. A rocket moves along the x-axis in S, commencing its motion with velocity v_0 and ending it with velocity v_1. If w is the jet velocity as measured by the crew (assumed constant), show that the mass ratio of the manoeuvre (i.e. initial mass/final mass) as measured by the crew is

$$\left[\frac{(c+v_1)(c-v_0)}{(c-v_1)(c+v_0)}\right]^{c/2w}$$

What does this reduce to as $c \to \infty$? Deduce that, if the rocket starts from rest in S and its jet is a stream of photons, the mass ratio to velocity v is

$$\sqrt{\left(\frac{c+v}{c-v}\right)}.$$

Show that, with a mass ratio of 6, the rocket can attain 35/37 of the velocity of light in S.

11. S, $\bar{S}$, $\bar{\bar{S}}$ are inertial frames with their axes parallel. $\bar{\bar{S}}$ has a velocity u relative to $\bar{S}$ and $\bar{S}$ has a velocity v relative to S, both velocities being parallel to the x-axes. If transformation from $\bar{S}$ to $\bar{\bar{S}}$ involves a rotation through an angle α of the axes in space-time and transformation from S to $\bar{S}$ a rotation β, a transformation from S to $\bar{\bar{S}}$ involves a rotation γ where $\gamma = \alpha + \beta$. Deduce from this equation the relativistic law for the composition of velocities, viz.

$$w = \frac{u+v}{1 + uv/c^2}$$

12. A force **f** acts upon a particle of mass m whose velocity is **v**. Show that

$$\mathbf{f} = m\frac{d\mathbf{v}}{dt} + \frac{\mathbf{f}\cdot\mathbf{v}}{c^2}\mathbf{v}.$$

13. An electrified particle having charge e and rest mass m_0 moves in a uniform electric field of intensity E parallel to the x-axis. If it is initially at rest at the origin, show that it moves along the x-axis so that at time t

$$x = \frac{c^2}{k}\left\{\sqrt{\left(1 + \frac{k^2}{c^2}t^2\right)} - 1\right\},$$

where $k = eE/m_0$. Show that this motion approaches that predicted by classical mechanics as $c \to \infty$. [It may be assumed that the force acting upon the particle is eE in the direction of the field at all times.]

14. A particle is moving with velocity u when it collides with a stationary particle having the same rest mass. After the collision the particles are moving at angles θ, ϕ with the direction of motion of the first particle before collision. Show that

$$\tan\theta\tan\phi = \frac{2}{\gamma + 1}$$

where $\gamma = (1 - u^2/c^2)^{-1/2}$. (If $c \to \infty$, $\gamma \to 1$ and $\theta + \phi = \tfrac{1}{2}\pi$. This is the prediction of classical mechanics. However, if the particles are electrons and u is near to c in value, $\theta + \phi < \tfrac{1}{2}\pi$. This effect has been observed in a Wilson cloud chamber.) [Hint: Refer the collision to an inertial frame in which both particles have equal and opposite velocities prior to collision.]

15. A body of mass M disintegrates while at rest into two parts of rest masses M_1 and M_2. Show that the energies E_1, E_2 of the parts are given by

$$E_1 = c^2\frac{M^2 + M_1^2 - M_2^2}{2M}, \quad E_2 = c^2\frac{M^2 - M_1^2 + M_2^2}{2M}.$$

16. A particle of proper mass m_0 moves under the action of a central force. (r, θ) are its polar coordinates in its plane of motion relative to the force centre as pole. $V(r)$ is its potential energy when at a distance r

from the centre. Obtain Lagrange's equations for the motion in the form

$$\frac{d}{dt}(\gamma \dot{r}) - \gamma r \dot{\theta}^2 + \frac{1}{m_0} V' = 0, \quad \frac{d}{dt}(\gamma r^2 \dot{\theta}) = 0,$$

where $\gamma = [1 - (\dot{r}^2 + r^2 \dot{\theta}^2)/c^2]^{-1/2}$. Write down the energy equation for the motion and obtain the differential equation for the orbit in the form

$$h^2 u^2 \left(\frac{d^2 u}{d\theta^2} + u \right) = \frac{C - V}{m_0^2 c^2} V',$$

where $u = 1/r$ and h, C are constants. In the inverse square law case when $V = -\mu/r$, deduce that the polar equation of the orbit can be written

$$lu = 1 + e \cos \eta \theta,$$

where $\eta^2 = 1 - \mu^2/m_0^2 h^2 c^2$. If $\mu/m_0 hc$ is small, show that the orbit is approximately an ellipse whose major axis rotates through an angle $\pi \mu^2/m_0^2 h^2 c^2$ per revolution.

17. A photon having energy E collides with a stationary electron whose rest mass is m_0. As a result of the collision the direction of the photon's motion is deflected through an angle θ and its energy is reduced to E'. Prove that

$$m_0 c^2 \left(\frac{1}{E'} - \frac{1}{E} \right) = 1 - \cos \theta.$$

(It may be assumed that the momentum of a photon having energy E is E/c.)

Deduce that the wavelength λ of the photon is increased by

$$\Delta \lambda = \frac{2h}{m_0 c} \sin^2 \tfrac{1}{2} \theta,$$

where h is Planck's constant. (This is the *Compton Effect*. For a photon, take $\lambda = hc/E$.)

18. A particle of rest mass m_1 and speed v collides with a particle of rest mass m_2 which is stationary. After collision the two particles

coalesce. Assuming that there is no radiation of energy, show that the rest mass of the combined particle is M, where

$$M^2 = m_1^2 + m_2^2 + \frac{2m_1 m_2}{(1-v^2/c^2)^{1/2}}$$

and find its speed.

19. A luminous disc of radius a has its centre fixed at the point $(\bar{x}, 0, 0)$ of the $\bar{S}$-frame and its plane is perpendicular to the $\bar{x}$-axis. It is observed from the origin in the S-frame at the instant the origins of the two frames coincide and is measured to subtend an angle 2α. Prove that, if $a \ll \bar{x}$, then

$$\tan \alpha = \frac{a}{\bar{x}}\sqrt{\left(\frac{c+u}{c-u}\right)}.$$

(Hint: employ the aberration of light formula, exercise 6 above.)

20. Two particles having proper masses m_1, m_2 are moving with velocities u_1, u_2 respectively, when they collide and cohere. If α is the angle between their lines of motion before collision, show that the proper mass of the combined particle is m, where

$$m^2 = m_1^2 + m_2^2 + \frac{2m_1 m_2(c^2 - u_1 u_2 \cos \alpha)}{\sqrt{\{(c^2 - u_1^2)(c^2 - u_2^2)\}}}.$$

Show that, for all values of α, $m \geq m_1 + m_2$ and explain the increase in proper mass.

21. $\mathbf{v}$, $\bar{\mathbf{v}}$ are the velocities of a point relative to the inertial frames S, $\bar{S}$ respectively. Representing these vectors as position vectors in an independent $\mathscr{E}_3$, show that

$$\beta \bar{\mathbf{v}} = Q\left[\mathbf{v} + \mathbf{u}\left\{\frac{\mathbf{u} \cdot \mathbf{v}}{u^2}(\beta - 1) + \beta\right\}\right],$$

where $\beta = (1 - u^2/c^2)^{-1/2}$ and

$$Q = 1/(1 + \mathbf{u} \cdot \mathbf{v}/c^2).$$

Show further that

$$u^2 \beta \bar{\mathbf{v}} = Q[(1-\beta)\mathbf{u} \times (\mathbf{v} \times \mathbf{u}) + \beta u^2(\mathbf{u} + \mathbf{v})],$$

SPECIAL RELATIVITY MECHANICS

and hence verify that

$$\bar{v}^2 = Q^2[(\mathbf{u}+\mathbf{v})^2 - (\mathbf{v} \times \mathbf{u})^2/c^2].$$

22. A particle moves along the x-axis of the frame S with velocity v and acceleration a. Show that the particle's acceleration in $\bar{S}$ is

$$\bar{a} = \frac{(1-u^2/c^2)^{3/2}}{(1-uv/c^2)^3} a.$$

If the particle always has constant acceleration α relative to an inertial frame in which it is instantaneously at rest, prove that

$$\frac{d}{dt}(\beta v) = \alpha,$$

where $\beta = (1-v^2/c^2)^{-1/2}$ and t is time in S.

Assuming that the particle is at rest at the origin of S at $t = 0$, show that its x-coordinate at time t is given by

$$\alpha x = c^2[(1+\alpha^2 t^2/c^2)^{1/2} - 1].$$

23. Three rectangular cartesian inertial frames S, $\bar{S}$, $\bar{\bar{S}}$ are initially coincident. As seen from S, $\bar{S}$ moves with velocity u parallel to Ox and, as seen from $\bar{S}$, $\bar{\bar{S}}$ moves with velocity v parallel to $\bar{O}\bar{y}$. If the direction of $\bar{\bar{S}}$'s motion as seen from S makes an angle θ with Ox and the direction of S's motion as seen from $\bar{\bar{S}}$ makes an angle ϕ with $\bar{\bar{O}}\bar{\bar{x}}$, prove that

$$\tan \theta = \frac{v}{u}\left(1-\frac{u^2}{c^2}\right)^{1/2}, \quad \tan \phi = \frac{v}{u}\left(1-\frac{v^2}{c^2}\right)^{-1/2}$$

Deduce that, if $u, v \ll c$, then

$$\phi - \theta = uv/2c^2$$

approximately.

24. A particle is moving with velocity v when it disintegrates into two photons having energies E_1, E_2, moving in directions making angles α, β with the original direction of motion and on opposite

sides of this direction. Show that

$$\tan \tfrac{1}{2}\alpha \tan \tfrac{1}{2}\beta = \frac{c-v}{c+v}.$$

Deduce that, if a photon disintegrates into two photons, they must both move in the same direction as the original photon. (The momentum of a photon having energy E is E/c.)

CHAPTER 4

Special Relativity Electrodynamics

21. 4-Current density

In this chapter we shall study the electromagnetic field due to a flow of charge which will be assumed known. Relative to an inertial frame S, let ρ be the charge density and $\mathbf{v}$ its velocity of flow. Then, if $\mathbf{j}$ is the current density,

$$\mathbf{j} = \rho \mathbf{v}. \tag{21.1}$$

Assuming that charge can neither be created nor destroyed, the equation of continuity

$$\text{div}\,\mathbf{j} + \frac{\partial \rho}{\partial t} = 0 \tag{21.2}$$

will be valid for the charge flow in S. This equation must be valid in every inertial frame and hence must be expressible in a form which is covariant with respect to orthogonal transformations in space-time. Introducing the coordinates x_i by equations (4.4) and employing equation (21.1), equation (21.2) is seen to be equivalent to

$$\frac{\partial}{\partial x_1}(\rho v_x) + \frac{\partial}{\partial x_2}(\rho v_y) + \frac{\partial}{\partial x_3}(\rho v_z) + \frac{\partial}{\partial x_4}(ic\rho) = 0. \tag{21.3}$$

This equation is covariant as required if $(\rho v_x, \rho v_y, \rho v_z, ic\rho)$ are the four components of a vector in space-time. For, if $\mathbf{J}$ is this vector, equation (21.3) can be written

$$J_{i,i} = 0, \tag{21.4}$$

and this is covariant with respect to orthogonal transformations. Now, by equation (15.6),

$$\mathbf{J} = (\rho \mathbf{v}, ic\rho) = \rho(1 - v^2/c^2)^{1/2}\mathbf{V}, \tag{21.5}$$

where **V** is the 4-velocity of flow and hence **J** is a vector if $\rho(1-v^2/c^2)^{1/2}$ is an invariant. Denoting the invariant by ρ_0, we have

$$\rho = \frac{\rho_0}{(1-v^2/c^2)^{1/2}}. \tag{21.6}$$

It follows that $\rho = \rho_0$ if $v = 0$ and hence that ρ_0 is the charge density as measured from an inertial frame relative to which the charge being considered is instantaneously at rest. ρ_0 is called the *proper charge density*.

J is called the *4-current density* and it is clear from equation (21.5) that

$$\mathbf{J} = \rho_0 \mathbf{V} = (\mathbf{j}, ic\rho). \tag{21.7}$$

It is now clear that, when **J** has been specified throughout space-time, the charge flow is completely determined, for the space components of **J** fix the current density and the time component fixes the charge density. Hence, given **J**, the electromagnetic field must be calculable. The equations which form the basis for this calculation will be derived in the next two sections.

Let $d\omega_0$ be the volume of a small element of charge as measured from an inertial frame S_0 relative to which the charge is instantaneously at rest. The total charge within the element is $\rho_0 d\omega_0$. Due to the Fitzgerald contraction, the volume of this element as measured from S will be $d\omega$, where

$$d\omega = (1-v^2/c^2)^{1/2} d\omega_0. \tag{21.8}$$

The total charge within the element as measured from S is therefore

$$\rho \, d\omega = \rho(1-v^2/c^2)^{1/2} d\omega_0 = \rho_0 d\omega_0, \tag{21.9}$$

by equation (21.6). It follows that the electric charge on a body is invariant for all inertial observers.

22. 4-Vector potential

In classical theory, the equations determining the electromagnetic field due to a given charge flow are Maxwell's equations (3.1)–(3.4). To ensure covariance of the laws of mechanics with respect to Lorentz transformations, it proved necessary to modify classical Newtonian theory slightly. However, it will be shown that Maxwell's equations

SPECIAL RELATIVITY ELECTRODYNAMICS

are covariant without any adjustment being necessary. Indeed, the Lorentz transformation equations were first noticed as the transformation equations which leave Maxwell's equations unaltered in form.

To prove this, it will be convenient to introduce the scalar and vector potentials, ϕ and $\mathbf{A}$ respectively, of the field. It is proved in textbooks devoted to the classical theory† that $\mathbf{A}$ satisfies the equations

$$\operatorname{div} \mathbf{A} + \frac{1}{c}\frac{\partial \phi}{\partial t} = 0, \tag{22.1}$$

$$\nabla^2 \mathbf{A} - \frac{1}{c^2}\frac{\partial^2 \mathbf{A}}{\partial t^2} = -\frac{4\pi}{c}\mathbf{j}, \tag{22.2}$$

and ϕ satisfies the equation

$$\nabla^2 \phi - \frac{1}{c^2}\frac{\partial^2 \phi}{\partial t^2} = -4\pi\rho. \tag{22.3}$$

We now define a 4-vector potential Ω in any inertial frame S by the equation

$$\Omega = (\mathbf{A}, i\phi). \tag{22.4}$$

It is easily verified that equations (22.2), (22.3) are together equivalent to the equation

$$\Box^2 \Omega = -\frac{4\pi}{c}\mathbf{J}, \tag{22.5}$$

where the operator $\Box^2$ is defined by

$$\Box^2 = \frac{\partial^2}{\partial x_1^2} + \frac{\partial^2}{\partial x_2^2} + \frac{\partial^2}{\partial x_3^2} + \frac{\partial^2}{\partial x_4^2}. \tag{22.6}$$

The space components of equation (22.5) yield equation (22.2) and the time component, equation (22.3). If Ω_i, J_i are the components of Ω and $\mathbf{J}$ respectively, equation (22.5) can be written

$$\Omega_{i,jj} = -\frac{4\pi}{c}J_i, \tag{22.7}$$

in which form it is clearly covariant with respect to Lorentz transformations provided Ω is a vector. This confirms that equation (22.4)

† See, e.g., *A Course in Applied Mathematics* by D. F. Lawden, p. 527. English Universities Press.

does, in fact, define a quantity with the transformation properties of a vector in space-time.

Next, it is necessary to show that equation (22.1) is also covariant with respect to orthogonal transformations in space-time. It is clearly equivalent to the equation

$$\operatorname{div} \boldsymbol{\Omega} = \Omega_{i,i} = 0, \tag{22.8}$$

which is in the required form.

J being given, $\boldsymbol{\Omega}$ is now determined by equations (22.7) and (22.8).

23. The field tensor

When **A** and ϕ are known in an inertial frame, the electric and magnetic intensities **E** and **H** respectively at any point in the electromagnetic field follow from the equations

$$\mathbf{E} = -\operatorname{grad} \phi - \frac{1}{c} \frac{\partial \mathbf{A}}{\partial t}, \tag{23.1}$$

$$\mathbf{H} = \operatorname{curl} \mathbf{A}. \tag{23.2}$$

Making use of equations (4.4) and (22.4), these equations are easily shown to be equivalent to the set

$$\left.\begin{aligned}
-iE_x &= \frac{\partial \Omega_4}{\partial x_1} - \frac{\partial \Omega_1}{\partial x_4}, \\
-iE_y &= \frac{\partial \Omega_4}{\partial x_2} - \frac{\partial \Omega_2}{\partial x_4}, \\
-iE_z &= \frac{\partial \Omega_4}{\partial x_3} - \frac{\partial \Omega_3}{\partial x_4},
\end{aligned}\right\} \tag{23.3}$$

$$\left.\begin{aligned}
H_x &= \frac{\partial \Omega_3}{\partial x_2} - \frac{\partial \Omega_2}{\partial x_3}, \\
H_y &= \frac{\partial \Omega_1}{\partial x_3} - \frac{\partial \Omega_3}{\partial x_1}, \\
H_z &= \frac{\partial \Omega_2}{\partial x_1} - \frac{\partial \Omega_1}{\partial x_2}.
\end{aligned}\right\} \tag{23.4}$$

Equations (23.3), (23.4) indicate that the six components of the vectors $-i\mathbf{E}, \mathbf{H}$ with respect to the rectangular Cartesian inertial frame S are the six distinct non-zero components in space-time of the skew-symmetric tensor $\Omega_{j,i} - \Omega_{,j}$. We have proved, therefore, that equations (23.1), (23.2) are valid in all inertial frames if

$$(F_{ij}) = \begin{pmatrix} 0 & H_z & -H_y & -iE_x \\ -H_z & 0 & H_x & -iE_y \\ H_y & -H_x & 0 & -iE_z \\ iE_x & iE_y & iE_z & 0 \end{pmatrix} \quad (23.5)$$

is assumed to transform as a tensor with respect to orthogonal transformations in space-time. The equations (23.3), (23.4) can then be summarized in the tensor equation

$$F_{ij} = \Omega_{j,i} - \Omega_{i,j}. \quad (23.6)$$

F_{ij} is called the *electromagnetic field tensor*. The close relationship between the electric and magnetic aspects of an electromagnetic field is now revealed as being due to their both contributing as components to the field tensor which serves to unite them.

Consider now equations (3.2), (3.3). Employing the field tensor defined by equation (23.5) and the current density given by equation (21.7), these equations are seen to be equivalent to

$$\left.\begin{aligned}
\frac{\partial F_{12}}{\partial x_2} + \frac{\partial F_{13}}{\partial x_3} + \frac{\partial F_{14}}{\partial x_4} &= \frac{4\pi}{c} J_1, \\
\frac{\partial F_{21}}{\partial x_1} + \frac{\partial F_{23}}{\partial x_3} + \frac{\partial F_{24}}{\partial x_4} &= \frac{4\pi}{c} J_2, \\
\frac{\partial F_{31}}{\partial x_1} + \frac{\partial F_{32}}{\partial x_2} + \frac{\partial F_{34}}{\partial x_4} &= \frac{4\pi}{c} J_3, \\
\frac{\partial F_{41}}{\partial x_1} + \frac{\partial F_{42}}{\partial x_2} + \frac{\partial F_{43}}{\partial x_3} &= \frac{4\pi}{c} J_4,
\end{aligned}\right\} \quad (23.7)$$

or, in short,

$$F_{ij,j} = \frac{4\pi}{c} J_i, \quad (23.8)$$

an equation which is covariant with respect to Lorentz transformations.

Finally, consider equations (3.1) and (3.4). These can be written

$$\left.\begin{aligned}
\frac{\partial F_{34}}{\partial x_2} + \frac{\partial F_{42}}{\partial x_3} + \frac{\partial F_{23}}{\partial x_4} &= 0, \\
\frac{\partial F_{41}}{\partial x_3} + \frac{\partial F_{13}}{\partial x_4} + \frac{\partial F_{34}}{\partial x_1} &= 0, \\
\frac{\partial F_{12}}{\partial x_4} + \frac{\partial F_{24}}{\partial x_1} + \frac{\partial F_{41}}{\partial x_2} &= 0, \\
\frac{\partial F_{23}}{\partial x_1} + \frac{\partial F_{31}}{\partial x_2} + \frac{\partial F_{12}}{\partial x_3} &= 0.
\end{aligned}\right\} \quad (23.9)$$

These equations are summarized thus:

$$F_{ij,k} + F_{jk,i} + F_{ki,j} = 0. \quad (23.10)$$

If any pair from i, j, k are equal, since F_{ij} is skew-symmetric, the left-hand member of this equation is identically zero and the equation is trivial. The four possible cases when i, j, k are distinct are the equations (23.9). Equation (23.10) is a tensor equation and is therefore also covariant with respect to Lorentz transformations.

To sum up, Maxwell's equations in 4-dimensional covariant form are:

$$\left.\begin{aligned}
F_{ij,j} &= \frac{4\pi}{c} J_i, \\
F_{ij,k} + F_{jk,i} + F_{ki,j} &= 0.
\end{aligned}\right\} \quad (23.11)$$

Given J_i at all points in space-time, these equations determine the field tensor F_{ij}. The solution can be found in terms of a vector potential Ω_i which satisfies the following equations:

$$\left.\begin{aligned}
\Omega_{i,i} &= 0, \\
\Omega_{i,jj} &= -\frac{4\pi}{c} J_i.
\end{aligned}\right\} \quad (23.12)$$

Ω_i being determined, F_{ij} follows from the equation

$$F_{ij} = \Omega_{j,i} - \Omega_{i,j}. \tag{23.13}$$

24. Lorentz transformations of electric and magnetic intensities

Since F_{ij} is a tensor, relative to the special Lorentz transformation (5.1) its non-zero components transform thus:

$$\left.\begin{aligned}\bar{F}_{23} &= F_{23}, \\ \bar{F}_{31} &= F_{31}\cos\alpha + F_{34}\sin\alpha, \\ \bar{F}_{12} &= F_{12}\cos\alpha + F_{42}\sin\alpha,\end{aligned}\right\} \tag{24.1}$$

$$\left.\begin{aligned}\bar{F}_{14} &= F_{14}, \\ \bar{F}_{24} &= -F_{21}\sin\alpha + F_{24}\cos\alpha, \\ \bar{F}_{34} &= -F_{31}\sin\alpha + F_{34}\cos\alpha.\end{aligned}\right\} \tag{24.2}$$

Substituting for the components of F_{ij} from equation (23.5) and for $\sin\alpha$, $\cos\alpha$ from equations (5.7), the above equations (24.1) yield the special Lorentz transformation equations for **H**, viz.

$$\bar{H}_x = H_x, \quad \bar{H}_y = \frac{H_y + (u/c)E_z}{(1-u^2/c^2)^{1/2}}, \quad \bar{H}_z = \frac{H_z - (u/c)E_y}{(1-u^2/c^2)^{1/2}}. \tag{24.3}$$

Similarly, equations (24.2) yield the transformation equations for **E**, viz.

$$\bar{E}_x = E_x, \quad \bar{E}_y = \frac{E_y - (u/c)H_z}{(1-u^2/c^2)^{1/2}}, \quad \bar{E}_z = \frac{E_z + (u/c)H_y}{(1-u^2/c^2)^{1/2}}. \tag{24.4}$$

As an example of the use to which these transformation formulae may be put, consider the electromagnetic field of a point charge e moving uniformly with speed u relative to the observer. Let S be the inertial frame employed by the observer and suppose e moves along its x-axis. $\bar{S}$ will be a parallel inertial frame with the point charge stationary at its origin. Consider the instant $t = 0$ in S, when the point charge is at the origin of S and the origins of S and $\bar{S}$ coincide. For an observer employing $\bar{S}$, the electromagnetic field is that due to a stationary point charge and is accordingly specified at the point $(\bar{x}, \bar{y}, \bar{z})$ for all $\bar{t}$ by the equations

$$\bar{\mathbf{E}} = \frac{e}{\bar{r}^3}(\bar{x}, \bar{y}, \bar{z}), \quad \bar{\mathbf{H}} = 0, \tag{24.5}$$

where $\bar{r}^2 = \bar{x}^2 + \bar{y}^2 + \bar{z}^2$. The field in S can now be calculated from the transformation equations inverse to (24.3), (24.4) (replace u by $-u$) and proves to be given by

$$\left.\begin{aligned}\mathbf{E} &= \frac{e}{\bar{r}^3}\left[\bar{x}, \frac{\bar{y}}{(1-u^2/c^2)^{1/2}}, \frac{\bar{z}}{(1-u^2/c^2)^{1/2}}\right], \\ \mathbf{H} &= \frac{e}{\bar{r}^3}\left[0, -\frac{u\bar{z}/c}{(1-u^2/c^2)^{1/2}}, \frac{u\bar{y}/c}{(1-u^2/c^2)^{1/2}}\right].\end{aligned}\right\} \tag{24.6}$$

But, putting $t = 0$ in equations (5.8), it will be found that

$$\bar{x} = \frac{x}{(1-u^2/c^2)^{1/2}}, \quad \bar{y} = y, \quad \bar{z} = z. \tag{24.7}$$

Whence, equations (24.6) are equivalent to

$$\left.\begin{aligned}\mathbf{E} &= \frac{e}{r'^3(1-u^2/c^2)^{1/2}}(x, y, z), \\ \mathbf{H} &= \frac{eu/c}{r'^3(1-u^2/c^2)^{1/2}}(0, -z, y),\end{aligned}\right\} \tag{24.8}$$

where $\quad r'^2 = \dfrac{x^2}{(1-u^2/c^2)} + y^2 + z^2.$

If $\mathbf{r}$ is the position vector of the point (x, y, z) with respect to the origin of S, equations (24.8) can be written

$$\left.\begin{aligned}\mathbf{E} &= \frac{e}{r'^3(1-u^2/c^2)^{1/2}}\mathbf{r}, \\ \mathbf{H} &= \frac{1}{c}\frac{e}{r'^3(1-u^2/c^2)^{1/2}}\mathbf{u}\times\mathbf{r}.\end{aligned}\right\} \tag{24.9}$$

SPECIAL RELATIVITY ELECTRODYNAMICS

These equations indicate that, at this instant in S, the **E**-lines are straight lines radiating from O and the **H**-lines are circles whose centres lie on Ox and whose planes are parallel to Oyz.

If $(u/c)^2$ is considered to be negligible, equations (24.9) reduce to

$$\mathbf{E} = \frac{e}{r^3}\mathbf{r}, \quad \mathbf{H} = \frac{1}{c}\frac{e}{r^3}\mathbf{u} \times \mathbf{r}. \tag{24.10}$$

The first equation shows that the electric field is, to this order of approximation, identical with the field of a stationary charge and the second equation is the *Biot-Savart Law*.

25. The Lorentz force

We shall now calculate the force exerted upon a point charge e in motion in an electromagnetic field.

At any instant, we can choose an inertial frame relative to which the point charge is instantaneously at rest. Let $\mathbf{E}_0$ be the electric intensity at the point charge relative to this frame. Then, by the physical definition of electric intensity as the force exerted upon unit stationary charge, the force exerted upon e will be $e\mathbf{E}_0$. It follows from equation (18.2) that the 4-force acting upon the charge in this frame is given by

$$\mathbf{F} = (e\mathbf{E}_0, 0). \tag{25.1}$$

The 4-velocity of the charge in this frame is also given by

$$\mathbf{V} = (0, ic) \tag{25.2}$$

and hence, by equation (23.5),

$$\frac{e}{c}F_{ij}V_j = e(E_{x0}, E_{y0}, E_{z0}, 0) = (e\mathbf{E}_0, 0). \tag{25.3}$$

It has accordingly been shown that, in an inertial frame relative to which the charge is instantaneously stationary,

$$F_i = \frac{e}{c}F_{ij}V_j. \tag{25.4}$$

But this is an equation between tensors and is therefore true for all inertial frames.

Substituting in equation (25.4) for the components F_i, F_{ij}, V_j from equations (18.2), (23.5) and (15.6) respectively, the following equations are obtained:

$$\left.\begin{aligned} f_x &= \frac{e}{c}(H_z v_y - H_y v_z + cE_x), \\ f_y &= \frac{e}{c}(H_x v_z - H_z v_x + cE_y), \\ f_z &= \frac{e}{c}(H_y v_x - H_x v_y + cE_z). \end{aligned}\right\} \quad (25.5)$$

These equations are equivalent to the 3-vector equation

$$\mathbf{f} = e\mathbf{E} + \frac{e}{c}\mathbf{v} \times \mathbf{H}. \quad (25.6)$$

f is called the *Lorentz force* acting upon the charged particle.

26. Force density

Consider a continuous distribution of matter in motion under the action of some field of force. Let $d\omega_0$ be the proper volume of any small element of the distribution and let **F** be the 4-force exerted upon the element by the field. Writing

$$\mathbf{F} = \mathbf{D}\, d\omega_0, \quad (26.1)$$

it follows that, since **F** is a vector and $d\omega_0$ is an invariant, then **D** is also a vector in space-time. It is termed the *4-force density* vector for the field.

When measured from an inertial frame S, let $d\omega$ be the volume of the element and let **f** be the 3-force exerted by the field upon it. We shall define the *3-force density* vector **d** in S by the equation

$$\mathbf{f} = \mathbf{d}\, d\omega. \quad (26.2)$$

$d\omega$ is an invariant with respect to transformations between Cartesian frames stationary in S and hence **d** is a vector relative to such frames.

SPECIAL RELATIVITY ELECTRODYNAMICS 69

Substituting for **F**, **f** from equations (26.1), (26.2) respectively into equation (18.2), we obtain

$$\mathbf{D}\,d\omega_0 = \left(\mathbf{d}, \frac{i}{c}\mathbf{d}\cdot\mathbf{v}\right) d\omega(1-v^2/c^2)^{-1/2}. \tag{26.3}$$

By virtue of the relationship (21.8), this reduces immediately to the equation

$$\mathbf{D} = \left(\mathbf{d}, \frac{i}{c}\mathbf{d}\cdot\mathbf{v}\right), \tag{26.4}$$

relating the 3- and 4-force density vectors.

27. The energy-momentum tensor for an electromagnetic field

Suppose that a charge distribution is specified by a 4-current density vector **J**. If $d\omega_0$ is the proper volume of any small element of the distribution and ρ_0 is the proper density of the charge, the charge within the element will be $\rho_0 d\omega_0$. It follows from equation (25.4) that the 4-force exerted upon the element by the electromagnetic field is given by

$$F_i = \frac{\rho_0}{c} F_{ij} V_j d\omega_0, \tag{27.1}$$

V being the 4-velocity of flow for the element. Employing equation (21.7), this last equation can be written

$$F_i = \frac{1}{c} F_{ij} J_j d\omega_0 \tag{27.2}$$

and it follows from the definition given in the last section that the 4-force density for the electromagnetic field is given by

$$D_i = \frac{1}{c} F_{ij} J_j. \tag{27.3}$$

Substituting for J_j from the first of equations (23.11), we can express D_i in terms of the field tensor thus:

$$D_i = \frac{1}{4\pi} F_{ij} F_{jk,k}. \tag{27.4}$$

F

We will now prove that the right-hand member of this equation is, apart from sign, the divergence of a certain symmetric tensor S_{ij} given by the equation

$$S_{ij} = \frac{1}{4\pi} F_{ik} F_{jk} - \frac{1}{16\pi} \delta_{ij} F_{kl} F_{kl} \tag{27.5}$$

and called the *energy-momentum tensor* of the electromagnetic field.

Taking the divergence of S_{ij}, we have

$$S_{ij,j} = \frac{1}{4\pi} F_{ik,j} F_{jk} + \frac{1}{4\pi} F_{ik} F_{jk,j} - \frac{1}{8\pi} \delta_{ij} F_{kl} F_{kl,j}. \tag{27.6}$$

Now

$$F_{ik,j} F_{jk} = F_{ij,k} F_{kj} = F_{ji,k} F_{jk}, \tag{27.7}$$

since F_{ij} is skew-symmetric. Thus

$$F_{ik,j} F_{jk} = \tfrac{1}{2}(F_{ik,j} + F_{ji,k}) F_{jk}. \tag{27.8}$$

Also

$$\delta_{ij} F_{kl} F_{kl,i} = F_{kl} F_{kl,i} = -F_{jk} F_{kj,i} \tag{27.9}$$

and it follows from these results that the first and last terms of the right-hand member of equation (27.6) can be combined to yield

$$\frac{1}{8\pi}(F_{ik,j} + F_{ji,k} + F_{kj,i}) F_{jk} \tag{27.10}$$

and this is zero by the second of equations (23.11).

Hence

$$S_{ij,j} = \frac{1}{4\pi} F_{ik} F_{jk,j} = -\frac{1}{4\pi} F_{ik} F_{kj,j} = -D_i. \tag{27.11}$$

Substituting for the components of the field tensor from equation (23.5), the components of S_{ij} are calculable from equation (27.5) as follows: If i, j take any of the values 1, 2, 3, then writing E_1 for E_x, E_2 for E_y, etc.

$$S_{ij} = -\frac{1}{4\pi}(E_i E_j + H_i H_j), \quad i \neq j. \tag{27.12}$$

If $i = j = 1$,

$$S_{11} = \frac{1}{4\pi}(H_2^2 + H_3^2 - E_1^2) - \frac{1}{8\pi}(H^2 - E^2),$$

$$= -\frac{1}{4\pi}(E_1^2 + H_1^2) + \frac{1}{8\pi}(E^2 + H^2). \qquad (27.13)$$

S_{22}, S_{33} may be expressed similarly and therefore, in general, if $i, j = 1, 2, 3$,

$$S_{ij} = -\frac{1}{4\pi}(E_i E_j + H_i H_j) + \frac{1}{8\pi}\delta_{ij}(E^2 + H^2). \qquad (27.14)$$

Apart from sign, this is *Maxwell's stress tensor* t_{ij}. t_{ij} is only a tensor with respect to rectangular frames stationary in the inertial frame being employed.

Also, if $i = 1, 2, 3$,

$$S_{i4} = S_{4i} = \frac{i}{4\pi}(E_2 H_3 - E_3 H_2, E_3 H_1 - E_1 H_3, E_1 H_2 - E_2 H_1)$$

$$= \frac{i}{4\pi} \mathbf{E} \times \mathbf{H} = \frac{i}{c}\mathbf{S}, \qquad (27.15)$$

where **S** is *Poynting's vector*.

Finally,

$$S_{44} = -\frac{1}{8\pi}(E^2 + H^2) = -U, \qquad (27.16)$$

where U is the energy density in the electromagnetic field.

These results may be summarized conveniently by exhibiting the components of S_{ij} in a matrix thus:

$$(S_{ij}) = -\left(\begin{array}{c|c} t_{ij} & \mathbf{S}/ic \\ \hline \mathbf{S}/ic & U \end{array}\right). \qquad (27.17)$$

Returning to the set of four equations (27.11), each can now be expressed in classical three-dimensional form. Thus, if $i = 1, 2, 3$, by reference to the equations (26.4) and (27.17) the corresponding equation is seen to be equivalent to

$$t_{ij,j} - \frac{1}{c^2}\frac{\partial S_i}{\partial t} = d_i. \qquad (27.18)$$

If $i = 4$, the corresponding equation is equivalent to

$$\operatorname{div}\mathbf{S} + \frac{\partial U}{\partial t} = -\mathbf{d}\cdot\mathbf{v}. \qquad (27.19)$$

Equations (27.18), (27.19) can be given simple physical interpretations in the case when the motion of a cloud of charged particles which do not interact mechanically is being studied. Each particle is then subjected to electromagnetic forces only and it will be assumed that its rest mass is constant throughout the motion, i.e. no heat is generated. Let Σ be a closed surface stationary in the frame S and let Γ be the region of space it encloses. Integrating equation (27.19) over Γ and employing Green's theorem, it will be found that

$$\int_\Sigma S_n\,d\sigma = \int_\Gamma \mathbf{d}\cdot\mathbf{v}\,d\omega + \frac{d}{dt}\int_\Gamma U\,d\omega, \qquad (27.20)$$

where $d\sigma$ is a surface element of Σ, $d\omega$ is a volume element of Γ and S_n is the component of **S** along the *inwards* normal to Σ. Now $\mathbf{d}\cdot\mathbf{v}\,d\omega$ is the rate at which the force applied to the charge in $d\omega$ is doing work and is therefore equal, by equation (17.7), to the rate of increase of the mechanical energy of this charge. The first term of the right-hand member of equation (27.20) accordingly gives the total rate at which the mechanical energy of the charge which is inside Σ at the instant under consideration is increasing. Since Σ is a fixed surface and the charge is moving, some of this mechanical energy will be lost by transport of charge across Σ. We have, therefore,

$$\int_\Gamma \mathbf{d}\cdot\mathbf{v}\,d\omega = \text{rate of increase of mechanical energy inside } \Sigma$$

$$+ \text{rate of loss of mechanical energy across } \Sigma. \quad (27.21)$$

The second term of the right-hand member of equation (27.20)

SPECIAL RELATIVITY ELECTRODYNAMICS 73

measures the rate of increase of electromagnetic field energy inside Σ. Equation (27.20) accordingly asserts that

$$\int_\Sigma S_n d\sigma + \text{rate of gain of mechanical energy across } \Sigma$$

$$= \text{total rate of energy increase within } \Sigma. \quad (27.22)$$

For the law of conservation of energy to be valid, it is clearly necessary to interpret the inward flux of **S** across Σ as the rate of flow of electromagnetic energy across this surface. Thus **S** is the *energy current density* vector.

Taking Σ to be a surface whose elements are all at a great distance from the charge flow being considered, so that $\mathbf{E} = 0$, $\mathbf{H} = 0$ over Σ and integrating the ith equation (27.18) over Γ, the contribution of the term $t_{ij,j}$ will be zero. This follows since $t_{ij,j}$ is the ordinary divergence of a vector having components (t_{i1}, t_{i2}, t_{i3}) and its volume integral, by Green's theorem, can be expressed as a surface integral of this vector over Σ and the vector is everywhere zero on Σ. There results, therefore,

$$\int_\Gamma d_i d\omega + \frac{d}{dt} \int_\Gamma \frac{1}{c^2} S_i d\omega = 0. \quad (27.23)$$

Now $d_i d\omega$ is the ith component of the force exerted upon the charge in $d\omega$ and therefore gives the rate at which its momentum is increasing. It follows from equation (27.23) that the net i^{th} component of momentum of an isolated system of charges is conserved only if the electromagnetic field is supposed to contribute momentum whose density is $\mathbf{S}/c^2$. This *electromagnetic momentum density* vector will be denoted by **g** and thus

$$\mathbf{g} = \frac{1}{c^2}\mathbf{S}. \quad (27.24)$$

If **w** is the velocity of propagation of electromagnetic energy, then

$$\mathbf{S} = U\mathbf{w} \quad (27.25)$$

and hence, by equation (27.24),

$$\mathbf{g} = \frac{U}{c^2}\mathbf{w}. \quad (27.26)$$

But, according to equation (17.14), U/c^2 is the mass density associated

74 TENSOR CALCULUS AND RELATIVITY

with an energy density U. It is consistent with earlier theory therefore, that such a mass density flowing with velocity **w** should generate a momentum density **g**.

28. Equations of motion of a charge flow

In this section, we shall continue to restrict our attention to a system comprising a cloud of charged particles which do not interact mechanically and whose proper masses remain constant during their motions.

Since the proper masses of the particles are conserved during the motion, an equation of continuity for proper mass can be found. Let Σ be any closed surface bounding a region Γ. Then the rate at which the net proper mass in Γ is decreasing must equal the rate at which proper mass is being lost by outwards flow across Σ. Let $d\omega$ be the volume of an element of the charge distribution as measured in the inertial frame S being employed and let $\mu_0 d\omega$ be the proper mass of the element. μ_0 is the *density of proper mass* in S. $\mu_0 d\omega$ is an invariant, but $d\omega$ is not and hence neither is μ_0. Then the mathematical expression of the statement we have just made is

$$-\frac{d}{dt}\int_{\Gamma} \mu_0 \, d\omega = \int_{\Sigma} \mu_0 v_n \, d\sigma, \qquad (28.1)$$

where v_n is the component of the velocity of flow **v** along the outward normal to Σ. Employing Green's theorem, equation (28.1) can be written

$$\int_{\Gamma} \left\{ \frac{\partial \mu_0}{\partial t} + \operatorname{div}(\mu_0 \mathbf{v}) \right\} d\omega = 0, \qquad (28.2)$$

and, since Γ is arbitrary, this implies the equation of continuity

$$\frac{\partial \mu_0}{\partial t} + \operatorname{div}(\mu_0 \mathbf{v}) = 0. \qquad (28.3)$$

Let $d\omega_0$ be the proper volume of a charge element and let $\mu_{00} d\omega_0$ be the proper mass of this element. Then μ_{00} is called the *proper density of proper mass*. Since $\mu_{00} d\omega_0$ and $d\omega_0$ are invariants, so is μ_{00}. Clearly

$$\mu_{00} d\omega_0 = \mu_0 d\omega \qquad (28.4)$$

SPECIAL RELATIVITY ELECTRODYNAMICS 75

and hence, by equation (21.8),

$$\mu_{00} = (1 - v^2/c^2)^{1/2} \mu_0. \tag{28.5}$$

It now follows that equation (28.3) can be written in the covariant form

$$\frac{\partial}{\partial x_i}(\mu_{00} V_i) = 0, \tag{28.6}$$

where V_i is the 4-velocity of flow.

The 4-force exerted by the field upon the charge element of proper volume $d\omega_0$ is, by equation (26.1), $D_i d\omega_0$. Since $\mu_{00} d\omega_0$ is the proper mass of this element, its equation of motion (see equation (17.2)) is therefore

$$D_i = \mu_{00} \frac{dV_i}{d\tau}. \tag{28.7}$$

But V_i can be expressed as a function of the x_i and, as the charge element moves, its coordinates x_i vary as functions of its proper time τ. It follows that

$$\frac{dV_i}{d\tau} = \frac{\partial V_i}{\partial x_j}\frac{dx_j}{d\tau} = \frac{\partial V_i}{\partial x_j} V_j. \tag{28.8}$$

Hence

$$\begin{aligned}
\frac{\partial}{\partial x_j}(\mu_{00} V_i V_j) &= \mu_{00} V_j \frac{\partial V_i}{\partial x_j} + V_i \frac{\partial}{\partial x_j}(\mu_{00} V_j), \\
&= \mu_{00} V_j \frac{\partial V_i}{\partial x_j}, \\
&= \mu_{00} \frac{dV_i}{d\tau}, \\
&= D_i, \tag{28.9}
\end{aligned}$$

equations (28.6), (28.8) and (28.7) having been employed to effect the reduction.

We now define a symmetric tensor Θ_{ij} by the equation

$$\Theta_{ij} = \mu_{00} V_i V_j, \tag{28.10}$$

so that D_i can be written as the divergence of this tensor thus

$$D_i = \Theta_{ij,j}. \qquad (28.11)$$

Consider the components of Θ_{ij}. We have

$$\Theta_{44} = -\frac{c^2\mu_{00}}{1-v^2/c^2} = -\frac{c^2\mu_0}{(1-v^2/c^2)^{1/2}} = -c^2\mu, \qquad (28.12)$$

where μ is the mass density in S. Hence, apart from sign, Θ_{44} is the mechanical energy density W.

If $i = 1, 2, 3$, then

$$\Theta_{i4} = \frac{ic\mu_{00}v_i}{1-v^2/c^2} = ic\mu v_i = icg_i, \qquad (28.13)$$

where $\mathbf{g} = \mu\mathbf{v}$ is the momentum density.

Finally, if $i, j = 1, 2, 3$,

$$\Theta_{ij} = \frac{\mu_{00}v_iv_j}{1-v^2/c^2} = \mu v_iv_j = g_iv_j. \qquad (28.14)$$

Now $g_i\mathbf{v}$ is the current density of the ith component of the momentum and thus the ith row of the 3×3 matrix (Θ_{ij}) can be so interpreted.

To summarize, we have shown that

$$(\Theta_{ij}) = \left(\begin{array}{c|c} g_iv_j & ic\mathbf{g} \\ \hline ic\mathbf{g} & -W \end{array}\right). \qquad (28.15)$$

This representation should be compared with the representation (27.17) of the electromagnetic energy-momentum tensor S_{ij}. Since, by equation (27.24), $\mathbf{S} = c^2\mathbf{g}$, it is clear that Θ_{ij} is the counterpart for the mass distribution of S_{ij} for the electromagnetic field. Θ_{ij} is called the *kinetic energy-momentum tensor*.

Substituting in the equation of motion (28.11) for D_i from equation (27.11), we find that

$$-S_{ij,j} = \Theta_{ij,j}. \qquad (28.16)$$

Writing

$$T_{ij} = S_{ij} + \Theta_{ij}, \qquad (28.17)$$

we have finally

$$T_{ij,j} = 0. \qquad (28.18)$$

T_{ij} is the overall energy-momentum tensor including contributions from an energy distribution in an electromagnetic form and from an energy distribution in a material form. Equation (28.18) indicates that the flow is determined by the statement that the divergence of the net energy-momentum tensor vanishes. This result can be proved to be true quite generally, i.e. corresponding to any energy (matter) distribution, there is an energy-momentum tensor T_{ij} whose divergence vanishes. It will be shown later in section 48 that this tensor also determines the gravitational field of the distribution.

As shown in the previous section, equation (28.18) with $i = 1, 2, 3$, expresses that the net linear momentum of the system is conserved and, with $i = 4$, that the total energy is conserved.

Exercises 4

1. Write down the special Lorentz transformation equations for **J** and deduce the transformation equations for **j**, ρ, viz.

$$\bar{j}_x = (1-u^2/c^2)^{-1/2}(j_x - \rho u), \qquad \bar{j}_y = j_y,$$
$$\bar{\rho} = (1-u^2/c^2)^{-1/2}(\rho - j_x u/c^2), \quad \bar{j}_z = j_z.$$

2. Deduce from the Maxwell equation

$$F_{ij,j} = \frac{4\pi}{c} J_i$$

that $\text{div}\, \mathbf{J} = 0$.

3. Verify that the field tensor defined in terms of the 4-potential Ω_i by equation (23.13) satisfies Maxwell's equations (23.11) provided Ω_i satisfies the equations (23.12).

4. (i) Prove that

$$F_{ij}F_{ij} = 2(H^2 - E^2)$$

and deduce that $H^2 - E^2$ is invariant with respect to Lorentz transformations.

(ii) Prove that

$$e_{ijkl} F_{ij} F_{kl} = -8i \mathbf{E} \cdot \mathbf{H}$$

and deduce that $\mathbf{E} \cdot \mathbf{H}$ is an invariant density with respect to Lorentz transformations.

5. An infinite uniform line charge lies along the x-axis of the inertial frame S and has longitudinal velocity $\mathbf{u}$. As measured in S, the charge per unit length is e. A point P is at distance r from the x-axis and $\mathbf{a}$ is the unit vector along the perpendicular to this axis through P. Show that the electric and magnetic field intensities at P are given by

$$\mathbf{E} = \frac{2e}{r}\mathbf{a}, \quad \mathbf{H} = \frac{2e}{cr}\mathbf{u}\times\mathbf{a}.$$

6. An observer O at rest in an inertial frame $Oxyzt$ finds himself to be in an electric field $\mathbf{E} = (0, E, 0)$, with no magnetic field. Show that an observer O' moving according to O with uniform velocity $\mathbf{V}$ at right angles to $\mathbf{E}$, finds electric and magnetic fields $\mathbf{E}'$, $\mathbf{H}'$ connected by the relation

$$c\,\mathbf{H}' + \mathbf{V}\times\mathbf{E}' = 0.$$

(L.U.)

7. *Oxyz* are rectangular axes. An electron moves in the xy-plane under the action of a uniform magnetic field parallel to Oz. Prove that its path is a circle.

8. A plane monochromatic electromagnetic wave is being propagated in a direction parallel to the x-axis in the inertial frame S. Its electric and magnetic field components are given by

$$\mathbf{E} = [0, a\sin\omega(t-x/c), 0],$$

$$\mathbf{H} = [0, 0, a\sin\omega(t-x/c)].$$

Show that, when observed from the inertial frame $\bar{S}$, it appears as the plane monochromatic wave

$$\bar{\mathbf{E}} = [0, \lambda a\sin\lambda\omega(\bar{t}-\bar{x}/c), 0],$$

$$\bar{\mathbf{H}} = [0, 0, \lambda a\sin\lambda\omega(\bar{t}-\bar{x}/c)],$$

where
$$\lambda = \sqrt{\left(\frac{1-u/c}{1+u/c}\right)},$$

u being the velocity of $\bar{S}$ relative to S. (I.e. both the amplitude and frequency are reduced by a factor λ. The reduction in frequency is the *Doppler effect*.)

9. Show that the Hamiltonian for the motion of a particle with charge e and mass m in an electromagnetic field $(\mathbf{A}, \phi)$ is

$$H = c\left[\left(\mathbf{p} - \frac{e}{c}\mathbf{A}\right)^2 + m^2 c^2\right]^{1/2} + e\phi$$

and express this equation in the covariant form

$$\left(\mathbf{P} - \frac{e}{c}\mathbf{\Omega}\right)^2 = -m^2 c^2.$$

10. Verify that, in a region devoid of charge, equations (23.12) are satisfied by

$$\Omega_i = A_i e^{ik_p x_p},$$

provided A_i, k_p are constants such that

$$A_i k_i = 0, \quad k_p k_p = 0.$$

By considering the 4-vector property of Ω_i, deduce that A_i must transform as a 4-vector under Lorentz transformations. Deduce also that $k_p x_p$ is a scalar under such transformations and hence that k_p is a 4-vector.

A plane electromagnetic wave, whose direction of propagation is parallel to the plane Oxy and makes an angle α with Ox, is given by

$$\Omega_i = A_i e^{2\pi i \nu(x\cos\alpha + y\sin\alpha - ct)/c},$$

where ν is the frequency. The same wave observed from a parallel frame $\bar{O}\bar{x}\bar{y}\bar{z}$ moving with velocity u along Ox, has frequency $\bar{\nu}$ and direction of propagation making an angle $\bar{\alpha}$ with $\bar{O}\bar{x}$. By writing down the transformation equations for the vector k_p, prove that

$$\bar{\nu} = \frac{1 - \frac{u}{c}\cos\alpha}{(1 - v^2/c^2)^{1/2}}\nu, \quad \cos\bar{\alpha} = \frac{\cos\alpha - \frac{u}{c}}{1 - \frac{u}{c}\cos\alpha}.$$

11. A particle of mass m and charge e moves freely in a magnetic field with components $(0, 0, H/z)$ and there is no electric field. Show that m is constant during the motion and, by a suitable choice of the

initial conditions, prove that the motion of the particle is given by

$$x = at\sin(\lambda \log t),$$
$$y = at\cos(\lambda \log t),$$
$$z = kt,$$

where $\lambda = eH/mck$ and hence that the particle moves on the surface of the cone

$$k^2(x^2+y^2) = a^2 z^2.$$

CHAPTER 5

General Tensor Calculus. Riemannian Space

29. Generalized N-dimensional spaces

In Chapter 2 the theory of tensors was developed in an N-dimensional Euclidean space on the understanding that the coordinate frame being employed was always rectangular Cartesian. If x_i, $x_i + dx_i$ are the coordinates of two neighbouring points relative to such a frame, the 'distance' ds between them is given by the equation

$$ds^2 = dx_i dx_i. \tag{29.1}$$

If $\bar{x}_i$, $\bar{x}_i + d\bar{x}_i$ are the coordinates of the same points with respect to another rectangular Cartesian frame, then

$$ds^2 = d\bar{x}_i d\bar{x}_i \tag{29.2}$$

and it follows that the expression $dx_i dx_i$ is invariant with respect to a transformation of coordinates from one rectangular Cartesian frame to another. Such a transformation was termed orthogonal.

Now, even in $\mathscr{E}_3$, it is very often convenient to employ a coordinate frame which is not Cartesian. For example, spherical polar coordinates (r, θ, ϕ) are frequently introduced, these being related to rectangular Cartesian coordinates (x, y, z) by the equations

$$x = r\sin\theta\cos\phi, \quad y = r\sin\theta\sin\phi, \quad z = r\cos\theta. \tag{29.3}$$

In such coordinates, the expression for ds^2 will be found to be

$$\begin{aligned} ds^2 &= dx^2 + dy^2 + dz^2, \\ &= dr^2 + r^2 d\theta^2 + r^2 \sin^2\theta\, d\phi^2, \end{aligned} \tag{29.4}$$

and this is no longer of the simple form of equation (29.1). The coordinate transformation (29.3) is accordingly not orthogonal. In fact, it is not even linear, as was the most general coordinate transformation (8.1) considered in Chapter 2.

The spherical polar coordinate system is an example of a *curvilinear coordinate frame* in $\mathscr{E}_3$. Let (u, v, w) be quantities related to rectangular Cartesian coordinates (x, y, z) by equations

$$u = u(x, y, z), \quad v = v(x, y, z), \quad w = w(x, y, z), \quad (29.5)$$

such that, to each point there corresponds a unique triad of values of (u, v, w) and to each such triad there corresponds a unique point. Then a set of values of (u, v, w) will serve to identify a point in $\mathscr{E}_3$ and (u, v, w) can be employed as coordinates. Such generalized coordinates are called *curvilinear coordinates*.

The equation

$$u(x, y, z) = u_0, \quad (29.6)$$

where u_0 is some constant, defines a surface in $\mathscr{E}_3$ over which u takes the constant value u_0. Similarly, the equations

$$v = v_0, \quad w = w_0 \quad (29.7)$$

define a pair of surfaces on which v takes the value v_0 and w the value w_0 respectively. These three surfaces will all pass through the point P_0 having coordinates (u_0, v_0, w_0) as shown in Fig. 4. They are called the *coordinate surfaces* through P_0. The surfaces $v = v_0, w = w_0$ will intersect in a curve $P_0 U$ along which v and w will be constant in value and only u will vary. $P_0 U$ is a *coordinate line* through P_0. Altogether, three coordinate lines pass through P_0. The equations $u = $ constant, $v = $ constant, $w = $ constant define three families of coordinate surfaces corresponding to the three families of planes parallel to the coordinate planes $x = 0, y = 0, z = 0$ of a rectangular Cartesian frame. Pairs of these surfaces intersect in coordinate lines which correspond to the parallels to the coordinate axes in a Cartesian frame.

Solving equations (29.5) for (x, y, z) in terms of (u, v, w), we obtain the inverse transformation

$$x = x(u, v, w), \quad y = y(u, v, w), \quad z = z(u, v, w). \quad (29.8)$$

Let (x, y, z), $(x + dx, y + dy, z + dz)$ be the rectangular Cartesian coordinates of two neighbouring points and let (u, v, w), $(u + du, v + dv, w + dw)$ be their respective curvilinear coordinates. Differentiating equations (29.8), we obtain

$$dx = \frac{\partial x}{\partial u} du + \frac{\partial x}{\partial v} dv + \frac{\partial x}{\partial w} dw, \text{ etc.} \quad (29.9)$$

GENERAL TENSOR CALCULUS. RIEMANNIAN SPACE 83

Thus, if ds is the distance between these points,

$$ds^2 = dx^2 + dy^2 + dz^2,$$
$$= A\,du^2 + B\,dv^2 + C\,dw^2 + 2F\,dv\,dw + 2G\,dw\,du + 2H\,du\,dv, \quad (29.10)$$

giving the appropriate expression for ds^2 in curvilinear coordinates. It will be noted that the coefficients A, B, etc. are, in general, functions of (u, v, w).

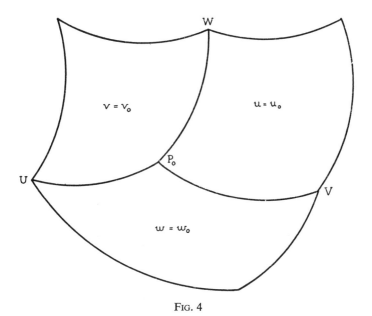

FIG. 4

If, therefore, curvilinear coordinate frames are to be permitted, the theory of tensors developed in Chapter 2 must be modified to make it independent of the special orthogonal transformations for which ds^2 is always expressible in the simple form of equation (29.1). The necessary modifications will be described in the later sections of this chapter. However, these modifications prove to be of such a nature that the amended theory makes no appeal to the special metrical properties of Euclidean space, i.e. the theory proves to be applicable in more general spaces for which Euclidean space is a particular case. This we shall now explain further.

Let $(x^1, x^2, \ldots, x^N)$ be curvilinear coordinates in $\mathscr{E}_N$.† Then, by analogy with equation (29.10), if ds is the distance between two neighbouring points, it can be shown that

$$ds^2 = g_{ij} dx^i dx^j, \qquad (29.11)$$

where the coefficients g_{ij} of the quadratic form in the x^i will, in general, be functions of these coordinates. Since the space is Euclidean, it is possible to transform from the curvilinear coordinates x^i to Cartesian coordinates y^i so that

$$ds^2 = dy^i dy^i. \qquad (29.12)$$

Clearly, the reduction of ds^2 to this simple form is only possible because the functions g_{ij} satisfy certain conditions. Conversely, the satisfaction of these conditions by the g_{ij} will guarantee that coordinates y^i exist for which ds^2 takes the simple form (29.12) and hence that the space is Euclidean. However, in extending the theory of tensors to be applicable to curvilinear coordinate frames, we shall, at a certain stage, make use of the fact that ds^2 is expressible in the form (29.11), but no use will be made of the conditions satisfied by the coefficients g_{ij} which are a consequence of the space being Euclidean. It follows that the extended theory will be applicable in a hypothetical N-dimensional space for which the 'distance' ds between neighbouring points $x^i, x^i + dx^i$ is given by an equation (29.11) in which the g_{ij} are *arbitrary functions* of the x^i.* Such a space is said to be *Riemannian* and will be denoted by $\mathscr{R}_N$. $\mathscr{E}_N$ is a particular $\mathscr{R}_N$ for which the g_{ij} satisfy certain conditions. The right-hand member of equation (29.11) is termed the *metric* of the Riemannian space.

The surface of the Earth provides an example of an $\mathscr{R}_2$. If θ is the

† The coordinates are here distinguished by superscripts instead of subscripts for a reason which will be given later.

* Except that partial derivatives of the g_{ij} will be assumed to exist and to be continuous to any order required by the theory.

GENERAL TENSOR CALCULUS. RIEMANNIAN SPACE 85

co-latitude and ϕ is the longitude of any point on the Earth's surface, the distance ds between the points (θ, ϕ), $(\theta+d\theta, \phi+d\phi)$ is given by

$$ds^2 = R^2(d\theta^2 + \sin^2\theta \, d\phi^2), \tag{29.13}$$

where R is the earth's radius. For this space and coordinate frame, the g_{ij} take the form

$$g_{11} = R^2, \quad g_{12} = g_{21} = 0, \quad g_{22} = R^2 \sin^2\theta. \tag{29.14}$$

It is not possible to define other coordinates (x, y) in terms of which

$$ds^2 = dx^2 + dy^2 \tag{29.15}$$

over the whole surface, i.e. this $\mathscr{R}_2$ is not Euclidean. However, the surfaces of a right circular cylinder and cone are Euclidean; the proof is left as an exercise for the reader.

It will be proved in Chapter 6 that, in the presence of a gravitational field, space-time ceases to be Euclidean in Minkowski's sense and becomes an $\mathscr{R}_4$. This is our chief reason for considering such spaces. However, we can generalize the concept of the space in which our tensors are to be defined yet further. Until section 39 is reached, we shall make no further reference to the metric. This implies that the theory of tensors, as developed thus far, is applicable in a very general N-dimensional space in which it is assumed it is possible to set up a coordinate frame but which is not assumed to possess a metric. In such a hypothetical space, the distance between two points is not even defined. It will be referred to as $\mathscr{S}_N$. $\mathscr{R}_N$ is a particular $\mathscr{S}_N$ for which a metric is specified.

30. Contravariant and covariant tensors

Let x^i be the coordinates of a point P in $\mathscr{S}_N$ relative to a coordinate frame which is specified in some manner which does not concern us here. Let $\bar{x}^i$ be the coordinates of the same point with respect to another reference frame and let these two systems of coordinates be related by equations

$$\bar{x}^i = \bar{x}^i(x^1, x^2, \ldots, x^N). \tag{30.1}$$

Consider the neighbouring point P' having coordinates $x^i + dx^i$ in the first frame. Its coordinates in the second frame will be $\bar{x}^i + d\bar{x}^i$, where

$$d\bar{x}^i = \frac{\partial \bar{x}^i}{\partial x^j} dx^j, \tag{30.2}$$

and summation with respect to the index j is understood. The N quantities dx^i are taken to be the components of the *displacement vector* **PP'** referred to the first frame. The components of this vector referred to the second frame are, correspondingly, the $d\bar{x}^i$ and these are related to the components in the first frame by the transformation equation (30.2). Such a displacement vector is taken to be the prototype for all *contravariant vectors*.

Thus, A^i are said to be the components of a contravariant vector located at the point x^i, if the components of the vector in the 'barred' frame are given by the equation

$$\bar{A}^i = \frac{\partial \bar{x}^i}{\partial x^j} A^j. \tag{30.3}$$

It is important to observe that, whereas in Chapter 2 the coefficients a_{ij} occurring in the transformation equation (10.2) were not functions of the Cartesian coordinates x_i so that the vector **A** was not, necessarily, located at a definite point in $\mathscr{E}_N$, the coefficients $\partial \bar{x}^i/\partial x^j$ in the corresponding equation (30.3) are functions of the x^i and the precise location of the vector A^i must be known before its transformation equations are determinate. This can be expressed by saying that there are no *free vectors* in $\mathscr{S}_N$.

The form of the transformation equation (30.3) should be studied carefully. It will be observed that the dummy index j occurs once as a superscript and once as a subscript (i.e. in the denominator of the partial derivative). Dummy indices will invariably occupy such positions in all expressions with which we shall be concerned in the sequel. Again, the free index i occurs as a superscript on both sides of the equation. This rule will be followed in all later developments, i.e. a free index will always occur in the same position (upper or lower) in each term of an equation. Finally, it will assist the reader to memorize this transformation if he notes that the free index is associated with the 'barred' symbol on both sides of the equation.

GENERAL TENSOR CALCULUS. RIEMANNIAN SPACE   87

A contravariant vector A^i may be defined at one point of $\mathscr{S}_N$ only. However, if it is defined at every point of a certain region, so that the A^i are functions of the x^i, a *contravariant vector field* is said to exist in the region.

If V is a quantity which is unaltered in value when the reference frame is changed, it is said to be a *scalar* or an *invariant* in $\mathscr{S}_N$. Its transformation equation is simply

$$\bar{V} = V. \tag{30.4}$$

Since this equation involves no coefficients dependent upon the x^i, the possibility that V may be a free invariant exists. However, V is more often associated with a specific point in $\mathscr{S}_N$ and may be defined at all points of a region of $\mathscr{S}_N$, in which case it defines an *invariant field*. In the latter case

$$V = V(x^1, x^2, \ldots, x^N). \tag{30.5}$$

$\bar{V}$ will then, in general, be a quite distinct function of the $\bar{x}^i$. If, however, in this function we substitute for the $\bar{x}^i$ in terms of the x^i from equation (30.1), by equation (30.4) the right-hand member of equation (30.5) must result. Thus

$$\bar{V}(\bar{x}^1, \bar{x}^2, \ldots, \bar{x}^N) = V(x^1, x^2, \ldots, x^N). \tag{30.6}$$

V being an invariant field, consider the N derivatives $\partial V/\partial x^i$. In the $\bar{x}^i$-frame, the corresponding quantities are $\partial \bar{V}/\partial \bar{x}^i$ and we have

$$\frac{\partial \bar{V}}{\partial \bar{x}^i} = \frac{\partial \bar{V}}{\partial x^j} \frac{\partial x^j}{\partial \bar{x}^i} = \frac{\partial x^j}{\partial \bar{x}^i} \frac{\partial V}{\partial x^j}, \tag{30.7}$$

since, by equation (30.6), when $\bar{V}$ is expressed as a function of the x it reduces to V. As in Chapter 2, the $\partial V/\partial x^i$ are taken to be the components of a vector called the *gradient* of V and denoted by grad V. However, its transformation law (30.7) is not the same as that for a contravariant vector, viz. (30.3) and it is taken to be the prototype for another species of vectors called *covariant vectors*.

Thus, B_i is a covariant vector if

$$\bar{B}_i = \frac{\partial x^j}{\partial \bar{x}^i} B_j. \tag{30.8}$$

Covariant vectors will be distinguished from contravariant vectors by writing their components with subscripts instead of superscripts. This notation is appropriate, for $\partial V/\partial x^i$ is a covariant vector and the index i occurs in the denominator of this partial derivative. The vector dx^i, on the other hand, has been shown to be contravariant in its transformation properties and this is correctly indicated by the upper position of the index. This is the reason for denoting the coordinates by x^i instead of x_i, although it must be clearly understood that the x^i alone are not the components of a vector at all.

The reader should check that the three rules formulated above in relation to the transformation equation (30.3), apply equally to the equation (30.8).

The generalization from vectors to tensors now proceeds along the same lines as in section 10. Thus, if A^i, B^j are two contravariant vectors, the N^2 quantities $A^i B^j$ are taken as the components of a contravariant tensor of the second rank. Its transformation equation is found to be

$$\bar{A}^i \bar{B}^j = \frac{\partial \bar{x}^i}{\partial x^k} \frac{\partial \bar{x}^j}{\partial x^l} A^k B^l. \tag{30.9}$$

Any set of N^2 quantities C^{ij} transforming in this way is a *contravariant tensor*.

Again, if A^i, B_j are vectors, the first contravariant and the second covariant, then the N^2 quantities $A^i B_j$ transform thus:

$$\bar{A}^i \bar{B}_j = \frac{\partial \bar{x}^i}{\partial x^k} \frac{\partial x^l}{\partial \bar{x}^j} A^k B_l. \tag{30.10}$$

Any set of N^2 quantities C^i_j transforming in this fashion is a *mixed tensor*, i.e. it possesses both contravariant and covariant properties as is indicated by the two positions of its indices.

Similarly, the transformation law for a covariant tensor of rank 2 can be assembled from the law for covariant vectors.

The further generalization to tensors of higher rank should now be an obvious step. It will be sufficient to give one example. A^i_{jk} is a mixed tensor of rank 3, having both the covariant and contravariant

GENERAL TENSOR CALCULUS. RIEMANNIAN SPACE 89

properties indicated by the positions of its indices, if it transforms according to the equation

$$\bar{A}^i_{jk} = \frac{\partial \bar{x}^i}{\partial x^r} \frac{\partial x^s}{\partial \bar{x}^j} \frac{\partial x^t}{\partial \bar{x}^k} A^r_{st}. \tag{30.11}$$

The components of a tensor can be given arbitrary values in any one frame and their values in any other frame are then uniquely determined by the transformation law. Consider the mixed second rank tensor whose components in the x^i-frame are δ^i_j, the Kronecker deltas ($\delta^i_j = 0$, $i \neq j$ and $\delta^i_j = 1$, $i = j$). The components in the $\bar{x}^i$-frame are $\bar{\delta}^i_j$, where

$$\bar{\delta}^i_j = \frac{\partial \bar{x}^i}{\partial x^k} \frac{\partial x^l}{\partial \bar{x}^j} \delta^k_l,$$

$$= \frac{\partial \bar{x}^i}{\partial x^k} \frac{\partial x^k}{\partial \bar{x}^j},$$

$$= \frac{\partial \bar{x}^i}{\partial \bar{x}^j},$$

$$= \delta^i_j. \tag{30.12}$$

Thus this tensor has the same components in all frames and is called the *fundamental mixed tensor*. However, a second rank *covariant* tensor whose components in the x^i-frame are the Kronecker deltas (in this case denoted by δ_{ij}), has different components in other frames and is accordingly of no special interest.

It is reasonable to enquire at this stage why the distinction between covariant and contravariant tensors did not arise when the coordinate transformations were restricted to be orthogonal. Thus, suppose that A^i, B_i are contravariant and covariant vectors with respect to the orthogonal transformation (8.1). The inverse transformation has been shown to be equation (11.5) and it follows from these two equations that

$$\frac{\partial \bar{x}_i}{\partial x_j} = a_{ij}, \quad \frac{\partial x_i}{\partial \bar{x}_j} = a_{ji}. \tag{30.13}$$

For the particular case of orthogonal transformations, therefore, equations (30.3), (30.8) take the form

$$\bar{A}^i = a_{ij} A^j, \quad \bar{B}_i = a_{ij} B_j. \tag{30.14}$$

It is clear that both types of vector transform in an identical manner and the distinction between them cannot, therefore, be maintained.

As in the case of the Cartesian tensors of Chapter 2, new tensors may be formed from known tensors by addition (or subtraction) and multiplication. Only tensors of the same rank and type may be added to yield new tensors. Thus, if A^i_{jk}, B^i_{jk} are components of tensors and we define the quantities C^i_{jk} by the equation

$$C^i_{jk} = A^i_{jk} + B^i_{jk}, \tag{30.15}$$

then C^i_{jk} are the components of a tensor having the covariant and contravariant properties indicated by the position of its indices. However, A^i_j, B_{ij} cannot be added in this way to yield a tensor. Any two tensors may be multiplied to yield a new tensor. Thus, if A^i_j, B^k_{lm} are tensors and we define N^5 quantities C^{ik}_{jlm} by the equation

$$C^{ik}_{jlm} = A^i_j B^k_{lm}, \tag{30.16}$$

these are the components of a fifth rank tensor having the covariant and contravariant properties indicated by its indices. The proofs of these statements are left for the reader to provide.

If a tensor is symmetric (or skew-symmetric) with respect to two of its superscripts or to two of its subscripts in any one frame, then it possesses this property in every frame. The method of proof is identical with that of the corresponding statement for Cartesian tensors given in section 10. However, if $A^i_j = A^j_i$ is true for all i, j when one reference frame is being employed, this equation will not, in general, be valid in any other frame. Thus, symmetry (or skew-symmetry) of a tensor with respect to a superscript and a subscript is not, in general, a covariant property. The tensor δ^i_j is exceptional in this respect.

Another result of great importance which may be established by the same argument we employed in the particular case of Cartesian tensors, is that an equation between tensors of the same type and rank is valid in all frames if it is valid in one. This implies that such tensor equations are covariant (i.e. are of invariable form) with respect to

GENERAL TENSOR CALCULUS. RIEMANNIAN SPACE 91

transformations between reference frames. The usefulness of tensors for our later work will be found to depend chiefly upon this property.

A symbol such as A^i_{jk} can be *contracted* by setting a superscript and a subscript to be the same letter. Thus A^i_{ji}, A^i_{ik} are the possible contractions of A^i_{jk} and each, by the repeated index summation convention, represents a sum. Since in the symbol A^i_{ji}, j alone is a free index, this entity has only N components. Similarly A^i_{ik} has N components. It will now be proved that, if A^i_{jk} is a tensor, its contractions are also tensors. Specifically, we shall prove that $B_j = A^i_{ji}$ is a covariant vector. For

$$\bar{B}_j = \bar{A}^i_{ji} = \frac{\partial \bar{x}^i}{\partial x^r} \frac{\partial x^s}{\partial \bar{x}^j} \frac{\partial x^t}{\partial \bar{x}^i} A^r_{st},$$

$$= \left(\frac{\partial x^t}{\partial \bar{x}^i} \frac{\partial \bar{x}^i}{\partial x^r}\right) \frac{\partial x^s}{\partial \bar{x}^j} A^r_{st},$$

$$= \frac{\partial x^t}{\partial x^r} \frac{\partial x^s}{\partial \bar{x}^j} A^r_{st},$$

$$= \delta^t_r \frac{\partial x^s}{\partial \bar{x}^j} A^r_{st},$$

$$= \frac{\partial x^s}{\partial \bar{x}^j} A^t_{st},$$

$$= \frac{\partial x^s}{\partial \bar{x}^j} B_s, \qquad (30.17)$$

establishing the result. This argument can obviously be generalized to yield the result that any contracted tensor is itself a tensor of rank two less than the tensor from which it has been derived and of the type indicated by the positions of its remaining free indices. In this connection it should be noted that, if A^i_{jk} is a tensor, A^i_{jj} is not, in general, a tensor; it is essential that the contraction be with respect to a superscript and a subscript and not with respect to two indices of the same kind.

If A^i_{jk}, B^r_s are tensors, the tensor $A^i_{jk} B^r_s$ is called the *outer product* of these two tensors. If this product is now contracted with respect to a superscript of one factor and a subscript of the other, e.g. $A^i_{jk} B^r_i$, the result is a tensor called an *inner product*.

31. The quotient theorem. Conjugate tensors

It has been remarked in the previous section that both the outer and inner products of two tensors are themselves tensors. Suppose, however, that it is known that a product of two factors is a tensor and that one of the factors is a tensor, can it be concluded that the other factor is also a tensor? We shall prove the following *quotient theorem*: *If the result of taking the product (outer or inner) of a given set of elements with a tensor of any specified type and arbitrary components is known to be a tensor, then the given elements are the components of a tensor.*

It will be sufficient to prove the theorem true for a particular case, since the argument will easily be seen to be of general application. Thus, suppose the A^i_{jk} are N^3 quantities and it is to be established that these are the components of a tensor of the type indicated by the positions of the indices. Let B^r_s be a mixed tensor of rank 2 whose components can be chosen arbitrarily (in any one frame only of course) and suppose it is given that the inner product

$$A^i_{jk} B^k_s = C^i_{js} \tag{31.1}$$

is a tensor for all such B^r_s. All components have been assumed calculated with respect to the x-frame. Transforming to the $\bar{x}$-frame, the inner product is given to transform as a tensor and hence we have

$$A^{i*}_{jk} \bar{B}^k_s = \bar{C}^i_{js}, \tag{31.2}$$

where A^{i*}_{jk} are the actual components replacing the A^i_{jk} when the reference frame is changed. Let $\bar{A}^i_{jk}$ be a set of elements *defined* in the $\bar{x}$-frame by equation (30.11). Since this is a tensor transformation equation, we know that the elements so defined will satisfy

$$\bar{A}^i_{jk} \bar{B}^k_s = \bar{C}^i_{js}. \tag{31.3}$$

Subtracting equation (31.3) from (31.2), we obtain

$$(A^{i*}_{jk} - \bar{A}^i_{jk}) \bar{B}^k_s = 0. \tag{31.4}$$

Since B^r_s has arbitrary components in the x-frame, its components in the $\bar{x}$-frame are also arbitrary and the components $\bar{B}^k_s$ can assume any

GENERAL TENSOR CALCULUS. RIEMANNIAN SPACE 93

convenient values. Thus, taking $\bar{B}_s^k = 1$ when $k = K$ and $\bar{B}_s^k = 0$ otherwise, equation (31.4) yields

$$A_{jK}^{i*} - \bar{A}_{jK}^i = 0,$$

or
$$A_{jK}^{i*} = \bar{A}_{jK}^i. \tag{31.5}$$

This being true for $K = 1, 2, \ldots, N$, we have quite generally

$$A_{jk}^{i*} = \bar{A}_{jk}^i. \tag{31.6}$$

This implies that A_{jk}^i does transform as a tensor.

We will first give a very simple example of the application of this theorem. Let A^i be an arbitrary contravariant vector. Then

$$\delta_j^i A^j = A^i \tag{31.7}$$

and since the right-hand member of this equation is certainly a vector, by the quotient theorem δ_j^i is a tensor (as we have proved earlier).

As a second example, let g_{ij} be a symmetric covariant tensor and let $g = |g_{ij}|$ be the determinant whose elements are the tensor's components. We shall denote by G^{ij} the co-factor in this determinant of the element g_{ij}. Then, although G^{ij} is not a tensor, if $g \neq 0$, $G^{ij}/g = g^{ij}$ is a symmetric contravariant tensor. To prove this, we first observe that

$$g_{ij} G^{kj} = g \delta_i^k, \quad g_{ij} G^{ik} = g \delta_j^k, \tag{31.8}$$

and hence, dividing by g,

$$g_{ij} g^{kj} = \delta_i^k, \quad g_{ij} g^{ik} = \delta_j^k. \tag{31.9}$$

Now let A^i be an arbitrary contravariant vector and define the covariant vector B_i by the equation

$$B_i = g_{ik} A^k. \tag{31.10}$$

Since $g \neq 0$, when the components of B_i are chosen arbitrarily, the corresponding components of A^i can always be calculated from this last equation, i.e. B_i is arbitrary with A^i. But

$$g^{ij} B_i = g^{ij} g_{ik} A^k = \delta_k^j A^k = A^j, \tag{31.11}$$

having employed the second identity (31.9). It now follows by the quotient theorem that g^{ij} is a contravariant tensor. That it is symmetric follows from the circumstance that G^{ij} possesses this property.

g_{ij}, g^{ij} are said to be *conjugate* to one another.

32. Relative tensors and tensor densities

The coordinates $\bar{x}^i$ being expressed in terms of the coordinates x^i by the equations (30.1), we shall define the determinant D of the transformation to have $\partial x^i/\partial \bar{x}^j$ for its ijth element. Thus

$$D = \left|\frac{\partial x^i}{\partial \bar{x}^j}\right|. \tag{32.1}$$

Then $\mathfrak{A}_j^i$ is a relative tensor of weight W (W a positive or negative integer) having both covariant and contravariant characteristics as are indicated by its indices, if its components transform thus:

$$\bar{\mathfrak{A}}_j^i = D^W \frac{\partial \bar{x}^l}{\partial x^r} \frac{\partial x^s}{\partial \bar{x}^j} \mathfrak{A}_s^r. \tag{32.2}$$

This statement can be generalized in the obvious way to relative tensors of any rank.

If $W = 0$, a relative tensor reduces to an ordinary tensor. In the particular case $W = 1$, the relative tensor is referred to as a *tensor density* (cf. section 13).

The following statements follow immediately from the definition of a relative tensor and the reader is left to devise formal proofs: (i) Relative tensors of the same type and weight can be added or subtracted to yield new tensors of the same type and weight; (ii) two relative tensors can be combined into an outer product and this will be a relative tensor whose weight is the sum of the weights of its factors; (iii) if a relative tensor is contracted with respect to a contravariant and a covariant index, its weight is unaltered but its rank is reduced by two.

$\mathfrak{e}^{ij\cdots n}$ denotes a contravariant tensor density of the Nth rank which is skew-symmetric with respect to every pair of indices. It then follows, as in section 13, that its components are determined as follows:

$\mathfrak{e}^{ij\cdots n} = 0$, if two indices are the same,

$\quad = +\mathfrak{e}^{12\cdots N}$, if $i, j, \ldots, n$ is an even permutation of $1, 2, \ldots, N$,

$\quad = -\mathfrak{e}^{12\cdots N}$, if $i, j, \ldots, n$ is an odd permutation of $1, 2, \ldots, N$.

GENERAL TENSOR CALCULUS. RIEMANNIAN SPACE

In particular, in the x-frame, we take $e^{12\ldots N} = 1$. Then, in the $\bar{x}$-frame,

$$\bar{e}^{12\ldots N} = D\frac{\partial \bar{x}^1}{\partial x^i}\frac{\partial \bar{x}^2}{\partial x^j}\cdots\frac{\partial \bar{x}^N}{\partial x^n}e^{ij\ldots n},$$

$$= DE, \tag{32.3}$$

where E is the determinant whose ij^{th} element is $\partial \bar{x}^i/\partial x^j$, i.e.

$$E = \left|\frac{\partial \bar{x}^i}{\partial x^j}\right|. \tag{32.4}$$

Employing the usual rule for the multiplication of determinants, from equations (32.1), (32.4) it follows that

$$DE = \left|\frac{\partial x^i}{\partial \bar{x}^r}\frac{\partial \bar{x}^r}{\partial x^j}\right| = \left|\frac{\partial x^i}{\partial x^j}\right| = |\delta^i_j| = 1. \tag{32.5}$$

Hence

$$\bar{e}^{12\ldots N} = 1, \tag{32.6}$$

proving that the components in the $\bar{x}$-frame are identical with those in the x-frame. The density $e^{ij\ldots n}$ accordingly possesses components $+1$, -1 and 0 in all frames.

It is left as an exercise for the reader to prove in a similar way that the skew-symmetric covariant relative tensor $e_{ij\ldots n}$ of the Nth rank and weight -1, whose component $e_{12\ldots N}$ is unity in the x-frame, has the same components $+1$, -1 and 0 in all frames.

If A_{ij} is a covariant tensor, the determinant $|A_{ij}|$ provides an example of a relative invariant. For its transformation law is

$$|\bar{A}_{ij}| = \left|\frac{\partial x^r}{\partial \bar{x}^i}A_{rs}\frac{\partial x^s}{\partial \bar{x}^j}\right|,$$

$$= \left|\frac{\partial x^r}{\partial \bar{x}^i}\right||A_{rs}|\left|\frac{\partial x^s}{\partial \bar{x}^j}\right|$$

$$= D^2|A_{rs}|. \tag{32.7}$$

Its weight is seen to be 2.

Similarly, if A^{ij} is a contravariant tensor, the determinant $|A^{ij}|$ will be found to transform according to the law

$$|\bar{A}^{ij}| = E^2|A^{ij}| = D^{-2}|A^{ij}| \tag{32.8}$$

and is therefore a relative invariant of weight -2.

Lastly, if A^i_j is a mixed tensor, the determinant $|A^i_j|$ is an invariant for its transformation law is:

$$|\bar{A}^i_j| = DE|A^i_j| = |A^i_j|. \tag{32.9}$$

33. Covariant derivatives. Parallel displacement. Affine connection

In the earlier sections of this chapter, the algebra of tensors was established and it is now time to explain how the concepts of analysis can be introduced into the theory. Our space $\mathscr{S}_N$ has N dimensions, but is otherwise almost devoid of special characteristics. Nonetheless, it has so far been able to provide all the facilities required of a stage upon which the tensors are to play their roles. It will now be demonstrated, however, that additional features must be built into the structure of $\mathscr{S}_N$, before it can function as a suitable environment for the operations of tensor analysis.

It has been proved that, if ϕ is an invariant field, $\partial \phi / \partial x^i$ is a covariant vector. But, if a covariant vector is differentiated, the result is not a tensor. For, let A_i be such a vector, so that

$$\bar{A}_i = \frac{\partial x^k}{\partial \bar{x}^i} A_k. \tag{33.1}$$

Differentiating both sides of this equation with respect to $\bar{x}^j$, we obtain

$$\frac{\partial \bar{A}_i}{\partial \bar{x}^j} = \frac{\partial x^k}{\partial \bar{x}^i} \frac{\partial x^l}{\partial \bar{x}^j} \frac{\partial A_k}{\partial x^l} + \frac{\partial^2 x^k}{\partial \bar{x}^i \partial \bar{x}^j} A_k. \tag{33.2}$$

The presence of the second term of the right-hand member of this equation reveals that $\partial A_i / \partial x^j$ does not transform as a tensor. However, this fact can be arrived at in a more revealing manner as follows:

Let P, P' be the neighbouring points x^i, $x^i + dx^i$ and let A_i, $A_i + dA_i$ be the vectors of a covariant vector field associated with these points respectively. The transformation laws for these two vectors will be

GENERAL TENSOR CALCULUS. RIEMANNIAN SPACE 97

different, since the coefficients of a tensor transformation law vary from point to point in $\mathscr{S}_N$. It follows that the difference of these two vectors, namely dA_i, is not a vector. However,

$$dA_i = \frac{\partial A_i}{\partial x^j} dx^j \tag{33.3}$$

and, since dx^j is a vector, if $A_{i,j}$ were a tensor, dA_i would be a vector. $A_{i,j}$ cannot be a tensor, therefore. The source of the difficulty is now apparent. To define $A_{i,j}$, it is necessary to compare the values assumed by the vector field A_i at two neighbouring, but distinct, points and such a comparison cannot lead to a tensor. If, however, this procedure could be replaced by another, involving the comparison of two vectors defined at the same point, the modified equation (33.3) would be expected to be a tensor equation featuring a new form of derivative which is a tensor. This leads us quite naturally to the concept of *parallel displacement*.

Suppose that the vector A_i is displaced from the point P, at which it is defined, to the neighbouring point P', *without change in magnitude or direction*, so that it may be thought of as being the same vector now defined at the neighbouring point. The phrase in italics has no precise meaning in $\mathscr{S}_N$ as yet, for we have not defined the magnitude or the direction of a vector in this space. However, in the particular case when $\mathscr{S}_N$ is Euclidean and rectangular axes are being employed, this phrase is, of course, interpreted as requiring that the displaced vector shall possess the same components as the original vector. But even in $\mathscr{E}_N$, if curvilinear coordinates are being used, the directions of the curvilinear axes at the point P' will, in general, be different from their directions at P and, as a consequence, the components of the displaced vector will not be identical with its components before the displacement. In $\mathscr{S}_N$, therefore, components of the displaced vector will be denoted by $A_i + \delta A_i$. This vector can now be compared with the field vector $A_i + dA_i$ at the same point P'. Since the two vectors are defined at the same point, their difference is a vector at this point, i.e. $dA_i - \delta A_i$ is a vector. The modified equation (33.3) is accordingly expected to be of the form

$$dA_i - \delta A_i = A_{i;j} dx^j, \tag{33.4}$$

where $A_{i;j}$ is the appropriate replacement for $A_{i,j}$. Since dx^j is an arbitrary vector and the left-hand member of equation (33.4) is known to be a vector, $A_{i;j}$ will, by the quotient theorem, be a covariant tensor. It will be termed the *covariant derivative* of A_i. Thus, the problem of defining a tensor derivative has been re-expressed as the problem of defining parallel displacement (infinitesimal) of a vector.

We are at liberty to define the parallel displacement of A_i from P to P' in any way we shall find convenient. However, to avoid confusion, it is necessary that the definition we accept shall be in conformity with that adopted in $\mathscr{E}_N$, which is a special case of $\mathscr{S}_N$. Suppose, therefore, that our $\mathscr{S}_N$ is Euclidean and that y are rectangular Cartesian coordinates in this space. Let B_i be the components of the vector field A_i with respect to these rectangular axes. Then

$$A_i = \frac{\partial y^j}{\partial x^i} B_j, \quad B_i = \frac{\partial x^j}{\partial y^i} A_j. \tag{33.5}$$

If the parallel displacement of the vector A_i to the point P' is now carried out, its Cartesian components B_i will not change, i.e. $\delta B = 0$. Hence, from the first of equations (33.5), we obtain

$$\delta A_i = \delta \left(\frac{\partial y^j}{\partial x^i} B_j \right) = \delta \left(\frac{\partial y^j}{\partial x^i} \right) B_j$$
$$= \frac{\partial^2 y^j}{\partial x^i \partial x^k} dx^k B_j. \tag{33.6}$$

Substituting for B_j into this equation from the second of equations (33.5), we find that

$$\delta A_i = \Gamma^l_{ik} A_l dx^k, \tag{33.7}$$

where

$$\Gamma^l_{ik} = \frac{\partial^2 y^j}{\partial x^i \partial x^k} \frac{\partial x^l}{\partial y^j}. \tag{33.8}$$

This shows that, in $\mathscr{E}_N$, the δA_i are bilinear forms in the A_l and dx^k. In $\mathscr{S}_N$, we shall accordingly *define* the δA_i by the equation (33.7), determining the N^3 quantities Γ^l_{ik} arbitrarily at every point of $\mathscr{S}_N$.†

† Subject to the requirement that the Γ^l_{ik} are continuous functions of the x^i and possess continuous partial derivatives to the order necessary to validate all later arguments.

GENERAL TENSOR CALCULUS. RIEMANNIAN SPACE   99

This set of quantities Γ^l_{ik} is called an *affinity* and specifies an *affine connection* between the points of $\mathscr{S}_N$. A space which is affinely connected possesses sufficient structure to permit the operations of tensor analysis to be carried out within it.

For we can now write

$$dA_i - \delta A_i = \frac{\partial A_i}{\partial x^j}dx^j - \Gamma^k_{ij}A_k\,dx^j,$$
$$= \left(\frac{\partial A_i}{\partial x^j} - \Gamma^k_{ij}A_k\right)dx^j. \qquad (33.9)$$

But, as we have already explained, the left-hand member of this equation is a vector for arbitrary dx^j and hence it follows that

$$A_{i;j} = \frac{\partial A_i}{\partial x^j} - \Gamma^k_{ij}A_k \qquad (33.10)$$

is a covariant tensor, the covariant derivative of A_i.

It will be observed from equation (33.10) that, if the components of the affinity all vanish over some region of $\mathscr{S}_N$, the covariant and partial derivatives are identical over this region. However, this will only be the case in the particular reference frame being employed. In any other frame the components of the affinity will, in general, be non-zero and the distinction between the two derivatives will be maintained. In tensor equations which are to be valid in every frame, therefore, only covariant derivatives may appear, even if it is possible to find a frame relative to which the affinity vanishes.

We have stated earlier that, when defining an affine connection, the components of an affinity may be chosen arbitrarily. To be precise, a coordinate frame must first be selected in $\mathscr{S}_N$ and the choice of the components of the affinity is then arbitrary within this frame. However, when these have been determined, the components of the affinity with respect to any other frame are, as for tensors, completely fixed by a transformation law. This transformation law for affinities, we now proceed to obtain.

34. Transformation of an Affinity

The manner in which each of the quantities occurring in equation (33.10) transforms is known, with the exception of the affinity Γ^k_{ij}. The transformation law for this affinity can accordingly be deduced by transformation of this equation. Relative to the $\bar{x}$-frame, the equation is written,

$$\bar{A}_{i;j} = \frac{\partial \bar{A}_i}{\partial \bar{x}^j} - \bar{\Gamma}^k_{ij} \bar{A}_k. \tag{34.1}$$

Since A_i, $A_{i;j}$ are tensors,

$$\bar{A}_i = \frac{\partial x^r}{\partial \bar{x}^i} A_r, \tag{34.2}$$

$$\bar{A}_{i;j} = \frac{\partial x^s}{\partial \bar{x}^i} \frac{\partial x^t}{\partial \bar{x}^j} A_{s;t}. \tag{34.3}$$

Substituting in equation (34.1), we obtain

$$\frac{\partial x^s}{\partial \bar{x}^i} \frac{\partial x^t}{\partial \bar{x}^j} A_{s;t} = \frac{\partial x^r}{\partial \bar{x}^i} \frac{\partial x^u}{\partial \bar{x}^j} \frac{\partial A_r}{\partial x^u} + \frac{\partial^2 x^r}{\partial \bar{x}^i \partial \bar{x}^j} A_r - \bar{\Gamma}^k_{ij} \frac{\partial x^r}{\partial \bar{x}^k} A_r. \tag{34.4}$$

Employing equation (33.10) to substitute for $A_{s;t}$ and cancelling a pair of identical terms from the two members of equation (34.4), this equation reduces to

$$-\frac{\partial x^s}{\partial \bar{x}^i} \frac{\partial x^t}{\partial \bar{x}^j} \Gamma^r_{st} A_r = \frac{\partial^2 x^r}{\partial \bar{x}^i \partial \bar{x}^j} A_r - \bar{\Gamma}^k_{ij} \frac{\partial x^r}{\partial \bar{x}^k} A_r. \tag{34.5}$$

Since A_r is an arbitrary vector, we can equate coefficients of A_r from the two members of this equation to obtain

$$\bar{\Gamma}^k_{ij} \frac{\partial x^r}{\partial \bar{x}^k} = \frac{\partial x^s}{\partial \bar{x}^i} \frac{\partial x^t}{\partial \bar{x}^j} \Gamma^r_{st} + \frac{\partial^2 x^r}{\partial \bar{x}^i \partial \bar{x}^j}. \tag{34.6}$$

Multiplying both sides of this equation by $\partial \bar{x}^l / \partial x^r$ and using the result

$$\frac{\partial \bar{x}^l}{\partial x^r} \frac{\partial x^r}{\partial \bar{x}^k} = \frac{\partial \bar{x}^l}{\partial \bar{x}^k} = \delta^l_k, \tag{34.7}$$

yields finally

$$\bar{\Gamma}^l_{ij} = \frac{\partial \bar{x}^l}{\partial x^r} \frac{\partial x^s}{\partial \bar{x}^i} \frac{\partial x^t}{\partial \bar{x}^j} \Gamma^r_{st} + \frac{\partial \bar{x}^l}{\partial x^r} \frac{\partial^2 x^r}{\partial \bar{x}^i \partial \bar{x}^j}, \tag{34.8}$$

which is the transformation law for an affinity.

It should be noted that, were it not for the presence of the second

GENERAL TENSOR CALCULUS. RIEMANNIAN SPACE

term in the right-hand member of equation (34.8), Γ^k_{ij} would transform as a tensor of the third rank having the covariant and contravariant characteristics suggested by the positions of its indices. Thus, the transformation law is linear in the components of an affinity but is not homogeneous like a tensor transformation law. This has the consequence that, if all the components of an affinity are zero relative to one frame, they are not necessarily zero relative to another frame. However, in general, there will be no frame in which the components of an affinity vanish over a region of $\mathscr{S}_N$, though it will be proved that, provided the affinity is symmetric, it is always possible to find a frame in which the components all vanish at some particular point (see section 38).

Suppose Γ^k_{ij}, Γ^{k*}_{ij} are two affinities defined over a region of $\mathscr{S}_N$. Writing down their transformation laws and subtracting one from the other, it is immediate that

$$\bar{\Gamma}^k_{ij} - \bar{\Gamma}^{k*}_{ij} = \frac{\partial \bar{x}^k}{\partial x^r} \frac{\partial x^s}{\partial \bar{x}^i} \frac{\partial x^t}{\partial \bar{x}^j} (\Gamma^r_{st} - \Gamma^{r*}_{st}), \tag{34.9}$$

i.e. the difference of two affinities is a tensor. However, the sum of two affinities is neither a tensor nor an affinity. It is left as an exercise for the reader to show, similarly, that the sum of an affinity Γ^k_{ij} and a tensor A^k_{ij} is an affinity.

If Γ^k_{ij} is symmetric with respect to its subscripts in one frame, it is symmetric in every frame. For, from equation (34.8),

$$\bar{\Gamma}^k_{ji} = \frac{\partial \bar{x}^k}{\partial x^r} \frac{\partial x^s}{\partial \bar{x}^j} \frac{\partial x^t}{\partial \bar{x}^i} \Gamma^r_{st} + \frac{\partial \bar{x}^k}{\partial x^r} \frac{\partial^2 x^r}{\partial \bar{x}^j \partial \bar{x}^i},$$

$$= \frac{\partial \bar{x}^k}{\partial x^r} \frac{\partial x^t}{\partial \bar{x}^i} \frac{\partial x^s}{\partial \bar{x}^j} \Gamma^r_{ts} + \frac{\partial \bar{x}^k}{\partial x^r} \frac{\partial^2 x^r}{\partial \bar{x}^i \partial \bar{x}^j},$$

$$= \bar{\Gamma}^k_{ij}, \tag{34.10}$$

where, at the first step, we have put $\Gamma^r_{ts} = \Gamma^r_{st}$.

35. Covariant derivatives of tensors

In this section, we shall extend the process of covariant differentiation to tensors of all ranks and types.

Consider first an invariant field V. When V suffers parallel displacement from P to P', its value will be taken to be unaltered, i.e. $\delta V = 0$ in all frames. Hence

$$dV - \delta V = \frac{\partial V}{\partial x^i} dx^i \qquad (35.1)$$

is the counterpart for an invariant of equation (33.4). It follows that

$$V_{;i} = V_{,i}, \qquad (35.2)$$

i.e. the covariant derivative of an invariant is identical with its partial derivative or gradient.

Now let B^i be a contravariant vector field and A_i an arbitrary covariant vector. Then $A_i B^i$ is an invariant and, when parallel displaced from P to P', remains unchanged in value. Thus

$$\delta(A_i B^i) = 0,$$

or
$$\delta A_i B^i + A_i \delta B^i = 0,$$

and hence, by equation (33.7),

$$A_k \delta B^k = -\Gamma^k_{ij} A_k dx^j B^i. \qquad (35.3)$$

But, since the A_k are arbitrary, their coefficients in the two members of this equation can be equated to yield

$$\delta B^k = -\Gamma^k_{ij} B^i dx^j. \qquad (35.4)$$

This equation defines the parallel displacement of a contravariant vector. The covariant derivative of the vector is now deduced as before: Thus

$$dB^k - \delta B^k = \left(\frac{\partial B^k}{\partial x^j} + \Gamma^k_{ij} B^i\right) dx^j \qquad (35.5)$$

and since dx^j is an arbitrary vector and $dB^k - \delta B^k$ is then known to be a vector,

$$B^k{}_{;j} = \frac{\partial B^k}{\partial x^j} + \Gamma^k_{ij} B^i \qquad (35.6)$$

is a tensor called the covariant derivative of B^k.

GENERAL TENSOR CALCULUS. RIEMANNIAN SPACE 103

Similarly, if A_j^i is a tensor field, we consider the parallel displacement of the invariant $A_j^i B_i C^j$, where B_i, C^j are arbitrary vectors. Then, from

$$\delta(A_j^i B_i C^j) = 0 \tag{35.7}$$

and equations (33.7), (35.4), we deduce that

$$\delta A_j^i = \Gamma_{jk}^l A_l^i dx^k - \Gamma_{lk}^i A_j^l dx^k. \tag{35.8}$$

It now follows that

$$A_{j;k}^i = \frac{\partial A_j^i}{\partial x^k} - \Gamma_{jk}^l A_l^i + \Gamma_{lk}^i A_j^l \tag{35.9}$$

is the covariant derivative required.

The rule for finding the covariant derivative of any tensor will now be plain from examination of equation (35.9), viz., the appropriate partial derivative is first written down and this is then followed by 'affinity terms'; the 'affinity terms' are obtained by writing down an inner product of the affinity and the tensor with respect to each of its indices in turn, prefixing a positive sign when the index is contravariant and a negative sign when it is covariant.

Applying this rule to the tensor field whose components at every point are those of the fundamental tensor δ_j^i, it will be found that

$$\delta_{j;k}^i = \Gamma_{rk}^i \delta_j^r - \Gamma_{jk}^r \delta_r^i = \Gamma_{jk}^i - \Gamma_{jk}^i = 0. \tag{35.10}$$

Thus, the fundamental tensor behaves like a constant in covariant differentiation.

Finally, in this section, we shall demonstrate that the ordinary rules for the differentiation of sums and products apply to the process of covariant differentiation.

The right-hand member of equation (35.9) being linear in the tensor A_j^i, it follows immediately that if

$$C_j^i = A_j^i + B_j^i, \tag{35.11}$$

then

$$C_{j;k}^i = A_{j;k}^i + B_{j;k}^i. \tag{35.12}$$

Now suppose that $$C^i = A_j^i B^j. \tag{35.13}$$

Then
$$C^i{}_{;k} = \frac{\partial C^i}{\partial x^k} + \Gamma^i_{rk} C^r,$$

$$= \frac{\partial}{\partial x^k}(A^i_j B^j) + \Gamma^i_{rk} A^r_j B^j,$$

$$= \left(\frac{\partial A^i_j}{\partial x^k} + \Gamma^i_{rk} A^r_j - \Gamma^r_{jk} A^i_r\right) B^j + \left(\frac{\partial B^j}{\partial x^k} + \Gamma^j_{rk} B^r\right) A^i_j,$$

$$= A^i_{j;k} B^j + B^j{}_{;k} A^i_j, \qquad (35.14)$$

which is the ordinary rule for the differentiation of a product.

36. Covariant differentiation of relative tensors

We consider first the parallel displacement of an invariant density $\mathfrak{A}$ from the point P to the neighbouring point P', at which it will be taken to become $\mathfrak{A} + \delta\mathfrak{A}$. It cannot be assumed that $\delta\mathfrak{A} = 0$ for, even if the two invariant densities are identical in one frame, since the transformation law at P is different from that at P', they will cease to be identical when referred to another frame. If, however, $\mathscr{S}_N$ is Euclidean and we restrict the choice of frames to those which are rectangular Cartesian, then the transformation law for the density will be

$$\bar{\mathfrak{A}} = D\mathfrak{A}, \qquad (36.1)$$

where D is given by equation (32.1) and, in the case of orthogonal transformations, can take the values $+1$ and -1 only. In this special case, therefore, D does not depend upon the point at which $\mathfrak{A}$ is defined and $\delta\mathfrak{A} = 0$ can be valid for all rectangular Cartesian frames. We shall accordingly define the parallel displacement of an invariant density in $\mathscr{E}_N$, relative to rectangular Cartesian frames, to be such that the density suffers no change.

Let y^i, $y^i + dy^i$ be the coordinates of points P, P' relative to rectangular Cartesian axes in $\mathscr{E}_N$ and let $\mathfrak{B}$ be an invariant density calculated in this frame and associated with the point P. If this density is parallel displaced to P', its value remains unchanged at $\mathfrak{B}$, i.e. $\delta\mathfrak{B} = 0$. Let x^i, $x^i + dx^i$ be the coordinates of P, P' respectively referred to any other coordinate frame (not necessarily Cartesian) and let $\mathfrak{A}$ be the

GENERAL TENSOR CALCULUS. RIEMANNIAN SPACE 105

same invariant density at P calculated in this new frame. Referred to the new frame, let $\mathfrak{A} + \delta\mathfrak{A}$ be this density after it has been parallel displaced to P'. Then

$$\mathfrak{A} = D\mathfrak{B}, \tag{36.2}$$

where $D = |\partial y^i/\partial x^j|$. Hence

$$\delta\mathfrak{A} = \delta(D\mathfrak{B}) = \delta D . \mathfrak{B} = \frac{1}{D}\delta D \mathfrak{A}. \tag{36.3}$$

But

$$\delta D = \frac{\partial D}{\partial x^i}dx^i \tag{36.4}$$

and thus

$$\delta\mathfrak{A} = K_i \mathfrak{A} dx^i, \tag{36.5}$$

where

$$K_i = \frac{1}{D}\frac{\partial D}{\partial x^i}. \tag{36.6}$$

Equation (36.5) is valid for the parallel displacement of an invariant density in a space $\mathscr{S}_N$ which is Euclidean. If $\mathscr{S}_N$ is not Euclidean, the parallel displacement of an invariant density will be *defined* by the equation (36.5). Since Cartesian coordinates are not available in such a space, the coefficients K_i cannot be derived from equation (36.6) and must, instead, be imposed upon the space by specification at every point (cf. an affinity). This specification will, for the moment, be assumed to be performed in an arbitrary manner in any one frame and, when this has been done, the K_i will be determined in all frames.

Suppose now that $\mathfrak{A}$ is an invariant density field and let $\mathfrak{A}$, $\mathfrak{A} + d\mathfrak{A}$ denote the actual values taken by the density at the points x^i, $x^i + dx^i$. Let $\mathfrak{A}$ be displaced to the point $x^i + dx^i$, where it assumes the value $\mathfrak{A} + \delta\mathfrak{A}$. Then the difference between the two densities defined at the point $x^i + dx^i$, is itself a density and equals

$$d\mathfrak{A} - \delta\mathfrak{A} = \left(\frac{\partial \mathfrak{A}}{\partial x^i} - K_i \mathfrak{A}\right)dx^i. \tag{36.7}$$

But dx^i being an arbitrary vector, it follows that

$$\mathfrak{A}_{;i} = \frac{\partial \mathfrak{A}}{\partial x^i} - K_i \mathfrak{A} \tag{36.8}$$

is a covariant vector density. This will be termed the *covariant derivative* of $\mathfrak{A}$.

Consider now $\mathfrak{B}_j^i$, a tensor density of the second rank. Let $\mathfrak{A}$ be any invariant density. Then $\mathfrak{A}^{-1}$ is a relative invariant of weight -1 and hence
$$\mathfrak{A}^{-1}\mathfrak{B}_j^i = B_j^i \tag{36.9}$$
is a tensor. Rewriting this equation thus,
$$\mathfrak{B}_j^i = \mathfrak{A} B_j^i, \tag{36.10}$$
and subjecting it to parallel displacement, we find
$$\begin{aligned}
\delta \mathfrak{B}_j^i &= \delta(\mathfrak{A} B_j^i), \\
&= \mathfrak{A}\, \delta B_j^i + B_j^i\, \delta \mathfrak{A}, \\
&= \mathfrak{A}(\Gamma_{jk}^r B_r^i - \Gamma_{rk}^i B_j^r)\, dx^k + B_j^i K_k \mathfrak{A} dx^k, \\
&= (\Gamma_{jk}^r \mathfrak{B}_r^i - \Gamma_{rk}^i \mathfrak{B}_j^r + K_k \mathfrak{B}_j^i)\, dx^k,
\end{aligned} \tag{36.11}$$
where equations (35.8) and (36.5) have been employed in the third line of this argument. Equation (36.11) indicates that the law of parallel displacement for a tensor density is identical with that for a tensor except for the presence of an additional term involving the coefficient K_k.

If $\mathfrak{B}_j^i$ is a tensor density field, its covariant derivative can now be found in the usual way. It is easily shown that this derivative is given by
$$\mathfrak{B}_{j;k}^i = \frac{\partial \mathfrak{B}_j^i}{\partial x^k} + \Gamma_{rk}^i \mathfrak{B}_j^r - \Gamma_{jk}^r \mathfrak{B}_r^i - K_k \mathfrak{B}_j^i \tag{36.12}$$
and is a tensor density. The rule for differentiating any tensor density should now be clear after inspection of the last equation.

Consider, in particular, the tensor density field whose components at every point are $e^{ij\ldots n}$. Since these components are all constant, their partial derivatives are zero and, by the rule, the covariant derivative is
$$e^{ij\ldots n}{}_{;s} = \Gamma_{rs}^i e^{rj\ldots n} + \Gamma_{rs}^j e^{ir\ldots n} + \ldots + \Gamma_{rs}^n e^{ij\ldots r} - K_s e^{ij\ldots n}. \tag{36.13}$$
If $i, j, \ldots, n$ is an even permutation of $1, 2, \ldots, N$, then
$$\left.\begin{aligned}
\Gamma_{rs}^i e^{rj\ldots n} &= \Gamma_{is}^i \quad \text{(not summed over } i\text{)}, \\
\Gamma_{rs}^j e^{ir\ldots n} &= \Gamma_{js}^j \quad \text{(not summed over } j\text{)},
\end{aligned}\right\} \tag{36.14}$$
etc. and hence
$$e^{ij\ldots n}{}_{;s} = \Gamma_{rs}^r - K_s, \tag{36.15}$$

GENERAL TENSOR CALCULUS. RIEMANNIAN SPACE 107

where summation with respect to r is intended. If $i, j, \ldots, n$ is an odd permutation, then

$$e^{ij\ldots n}{}_{;s} = -(\Gamma^r_{rs} - K_s). \tag{36.16}$$

If a pair of $i, j, \ldots n$ are the same, e.g. $i = j = P$, then

$$\begin{aligned}
\Gamma^i_{rs} e^{rj\ldots n} + \Gamma^j_{rs} e^{ir\ldots n} &= \Gamma^P_{rs} e^{rP\ldots n} + \Gamma^P_{rs} e^{Pr\ldots n}, \\
&= \Gamma^P_{rs} e^{rP\ldots n} - \Gamma^P_{rs} e^{rP\ldots n}, \\
&= 0, \tag{36.17}
\end{aligned}$$

where summation with respect to P is not intended. Also, the remaining terms of the right-hand member of equation (36.13) are then zero and it follows that, in this case,

$$e^{ij\ldots n}{}_{;s} = 0. \tag{36.18}$$

Equations (36.15), (36.16), (36.18) can be summarized in the form

$$e^{ij\ldots n}{}_{;s} = (\Gamma^r_{rs} - K_s) e^{ij\ldots n}. \tag{36.19}$$

$e^{ij\ldots n}$ having the same components in every frame and at every point, it is natural to expect its covariant derivative to be identically zero. Such a result would certainly facilitate calculations into which this density enters, since it could then be treated as a constant with respect to covariant differentiation (cf. δ^i_j). Equation (36.19) shows that this desirable state of affairs can easily be achieved by taking

$$K_s = \Gamma^r_{rs} \tag{36.20}$$

in the law defining parallel displacement of densities. This we shall accordingly do in all future developments. The covariant derivative of a tensor density as determined by equation (36.12), will then read

$$\mathfrak{B}^i_{j;k} = \frac{\partial \mathfrak{B}^i_j}{\partial x^k} + \Gamma^i_{rk} \mathfrak{B}^r_j - \Gamma^r_{jk} \mathfrak{B}^i_r - \Gamma^r_{rk} \mathfrak{B}^i_j. \tag{36.21}$$

The parallel displacement of any relative tensor of weight W can now be deduced. Let $\mathfrak{C}$ be a relative invariant of weight W and $\mathfrak{A}$ an invariant density. Then $\mathfrak{C}/\mathfrak{A}^W$ is an invariant. Denoting this by V, we have

$$\mathfrak{C} = \mathfrak{A}^W V. \tag{36.22}$$

Performing a parallel displacement to a neighbouring point, we obtain

$$\delta \mathfrak{C} = \delta(\mathfrak{A}^W V) = W \mathfrak{A}^{W-1} \delta \mathfrak{A} V, \tag{36.23}$$

since $\delta V = 0$. Substituting for $\delta \mathfrak{A}$ from equation (36.5) (using the above form for K_i), it is found that

$$\delta \mathfrak{C} = W \Gamma^i_{ik} \mathfrak{A}^W V dx^k = W \Gamma^i_{ik} \mathfrak{C} dx^k. \tag{36.24}$$

This is the law for parallel displacement of a relative invariant of weight W and it will be noted that it is identical with the law (36.5) for the parallel displacement of a density, except for the occurrence of the numerical factor W.

The derivation of the covariant derivatives of relative tensors now proceeds in the same manner as for densities. The covariant derivative of an invariant $\mathfrak{C}$ of weight W proves to be

$$\mathfrak{C}_{;k} = \frac{\partial \mathfrak{C}}{\partial x^k} - W \Gamma^i_{ik} \mathfrak{C}, \tag{36.25}$$

(cf. equation (36.8)), and

$$\mathfrak{C}^i_{j;k} = \frac{\partial \mathfrak{C}^i_j}{\partial x^k} + \Gamma^i_{rk} \mathfrak{C}^r_j - \Gamma^r_{jk} \mathfrak{C}^i_r - W \Gamma^r_{rk} \mathfrak{C}^i_j \tag{36.26}$$

is the covariant derivative of a relative tensor $\mathfrak{C}^i_j$ of weight W (cf. equation (36.21)).

The covariant relative tensor of weight -1, which has earlier been denoted by $\mathfrak{e}_{ij...n}$, can be regarded as a field and its derivative now found. It will be left as an exercise for the reader to show that this derivative is identically zero. The method used to derive equation (36.19) should be employed.

37. The Riemann-Christoffel curvature tensor

If a rectangular Cartesian coordinate frame is chosen in a Euclidean space $\mathscr{E}_N$ and if A^i are the components of a vector defined at a point Q with respect to this frame, then $\delta A^i = 0$ for an arbitrary small parallel displacement of the vector from Q. This being true for arbitrary A^i, it follows from equation (35.4) that $\Gamma^i_{jk} = 0$ with respect to this frame at every point of $\mathscr{E}_N$. Suppose C is a closed curve passing through Q and

GENERAL TENSOR CALCULUS. RIEMANNIAN SPACE 109

that A^i makes one complete circuit of C, being parallel displaced over each element of the path. Then its components remain unchanged throughout the motion and hence, if $A^i + \Delta A^i$ denotes the vector upon its return to Q,

$$\Delta A^i = 0. \tag{37.1}$$

Since ΔA^i is the difference between two vectors both defined at Q, it is itself a vector and equation (37.1) will therefore be a vector equation true in all frames. Thus, in $\mathscr{E}_N$, parallel displacement of a vector through one circuit of a closed curve leaves the vector unchanged.

If, however, A^i is defined at a point Q in an affinely connected space $\mathscr{S}_N$, not necessarily Euclidean, it will no longer be possible, in general, to choose a coordinate frame for which the components of the affinity vanish at every point. As a consequence, if A^i is parallel displaced around C, its components will vary and it is no longer permissible to suppose that upon its return to Q it will be unchanged, i.e. $\Delta A^i \neq 0$. We shall now calculate ΔA^i when A^i is parallel displaced around a small circuit C enclosing the point P having coordinates x^i (Fig. 5) at which it is initially defined.

Let U be any point on this curve and let $x^i + \xi^i$ be its coordinates, the ξ^i being small quantities. V is a point on C near to U and having coordinates $x^i + \xi^i + d\xi^i$. When A^i is displaced from U to V, its components undergo a change

$$\delta A^i = -\Gamma^i_{jk} A^j d\xi^k, \tag{37.2}$$

where Γ^i_{jk} and A^j are to be computed at U. Considering the small displacement from P to U and employing Taylor's theorem, the value of Γ^i_{jk} at U is seen to be

$$\Gamma^i_{jk} + \frac{\partial \Gamma^i_{jk}}{\partial x^l} \xi^l, \tag{37.3}$$

to the first order in the ξ^l. In this expression, the affinity and its derivative are to be computed at P. A^j in equation (37.2) represents the vector after its parallel displacement from P to U, i.e. it is

$$A^j - \Gamma^j_{rl} A^r \xi^l, \tag{37.4}$$

where A^j, A^r and Γ^j_{rl} are all to be calculated at the point P. To the first order in ξ^l therefore, equation (37.2) may be written

$$\delta A^i = -\left[\Gamma^i_{jk} A^j + \left(A^j \frac{\partial \Gamma^i_{jk}}{\partial x^l} - \Gamma^i_{jk}\Gamma^j_{rl} A^r\right)\xi^l\right] d\xi^k. \quad (37.5)$$

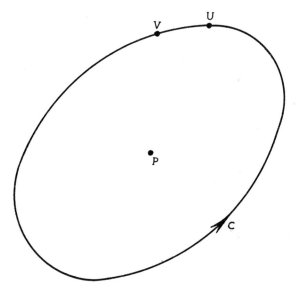

Fig. 5

Integrating around C, it will be found that

$$\Delta A^i = -\Gamma^i_{jk} A^j \oint_C d\xi^k + \left(\Gamma^i_{rk}\Gamma^r_{jl} - \frac{\partial \Gamma^i_{jk}}{\partial x^l}\right) A^j \oint_C \xi^l d\xi^k, \quad (37.6)$$

where the dummy indices j, r have been interchanged in the final term of the right-hand member of equation (37.5).

Now
$$\oint_C d\xi^k = \Delta \xi^k = 0. \quad (37.7)$$

GENERAL TENSOR CALCULUS. RIEMANNIAN SPACE 111

Also
$$\oint_C d(\xi^l \xi^k) = \Delta(\xi^l \xi^k) = 0, \tag{37.8}$$

so that
$$\oint_C \xi^l d\xi^k = -\oint_C \xi^k d\xi^l, \tag{37.9}$$

implying that the left-hand member of this equation is skew-symmetric in l and k. Since $\xi^l, d\xi^k$ are vectors, it is also a tensor. Denoting it by α^{kl}, we have

$$\alpha^{kl} = \tfrac{1}{2} \oint_C (\xi^l d\xi^k - \xi^k d\xi^l) \tag{37.10}$$

and equation (37.6) then reduces to the form

$$\Delta A^i = \left(\Gamma^i_{rk} \Gamma^r_{jl} - \frac{\partial \Gamma^i_{jk}}{\partial x^l} \right) A^j \alpha^k . \tag{37.11}$$

Apart from its property of skew-symmetry, α^{kl} is arbitrary. Nonetheless, since it is not completely arbitrary, the quotient theorem (section 31) cannot be applied directly to deduce that the contents of the bracket in equation (37.11) constitute a tensor. In fact, this expression is not a tensor. However, it is easy to prove that, if X^{ij}_{kl} is skew-symmetric with respect to k, l and if Y^{ij}, defined by the equation

$$Y^{ij} = X^{ij}_{kl} \alpha^{kl}, \tag{37.12}$$

is a tensor for arbitrary skew-symmetric tensors α^{kl}, then X^{ij}_{kl} is also a tensor.

To prove this, let β^{kl} be an arbitrary symmetric tensor. Then the components of the tensor

$$\gamma^{kl} = \alpha^{kl} + \beta^k \tag{37.13}$$

are completely arbitrary, for, assuming $k < l$,

$$\gamma^{kl} = \alpha^{kl} + \beta^{kl}, \quad \gamma^{lk} = -\alpha^{kl} + \beta^{kl} \tag{37.14}$$

and it follows that the values of γ^{kl}, γ^{lk} can be chosen arbitrarily and then

$$\alpha^{kl} = \tfrac{1}{2}(\gamma^{kl} - \gamma^{lk}), \quad \beta^{kl} = \tfrac{1}{2}(\gamma^{kl} + \gamma^{lk}). \tag{37.15}$$

I.e., it is only necessary to fix the values of α^{kl}, β^{kl} in the cases $k < l$ in order that the γ^{kl} shall assume any specified values over the complete range of its superscripts, with the exception of the cases when two superscripts are equal. If the superscripts are equal, $\alpha^{kl} = 0$ and $\gamma^{kl} = \beta^{kl}$. But these β^{kl} are also arbitrary and hence so again are the γ^{kl} with equal superscripts.

Since β^{kl} is symmetric and X_{kl}^{ij} is skew-symmetric,

$$X_{kl}^{ij}\beta^{kl} = 0. \qquad (37.16)$$

Adding equations (37.12) and (37.16), we obtain therefore

$$X_{kl}^{ij}\gamma^{kl} = Y^{ij}. \qquad (37.17)$$

But γ^{kl} is an arbitrary tensor and hence, by the quotient theorem, X_{kl}^{ij} is a tensor.

The multiplier of α^{kl} in equation (37.11) is not skew-symmetric in k, l. However, it can be made so as follows: Interchange the dummy indices k, l in this equation to obtain

$$\Delta A^i = \left(\Gamma_{rl}^i \Gamma_{jk}^r - \frac{\partial \Gamma_{jl}^i}{\partial x^k}\right) A^j \alpha^{lk}. \qquad (37.18)$$

Adding equations (37.11) and (37.18) and noting that $\alpha^{kl} = -\alpha^{lk}$, it will be found that

$$\Delta A^i = \tfrac{1}{2}\left(\Gamma_{rk}^i \Gamma_{jl}^r - \Gamma_{rl}^i \Gamma_{jk}^r + \frac{\partial \Gamma_{jl}^i}{\partial x^k} - \frac{\partial \Gamma_{jk}^i}{\partial x^l}\right) A^j \alpha^{kl}. \qquad (37.19)$$

The bracketed expression is now skew-symmetric in k, l and hence

$$\left(\Gamma_{rk}^i \Gamma_{jl}^r - \Gamma_{rl}^i \Gamma_{jk}^r + \frac{\partial \Gamma_{jl}^i}{\partial x^k} - \frac{\partial \Gamma_{jk}^i}{\partial x^l}\right) A^j \qquad (37.20)$$

is a tensor. A^j being arbitrary, it follows that

$$B_{jkl}^i = \Gamma_{rk}^i \Gamma_{jl}^r - \Gamma_{rl}^i \Gamma_{jk}^r + \frac{\partial \Gamma_{jl}^i}{\partial x^k} - \frac{\partial \Gamma_{jk}^i}{\partial x^l} \qquad (37.21)$$

is a tensor. It is the *Riemann-Christoffel Curvature Tensor*.

Equation (37.19) can now be written

$$\Delta A^i = \tfrac{1}{2} B_{jkl}^i A^j \alpha^{kl}. \qquad (37.22)$$

GENERAL TENSOR CALCULUS. RIEMANNIAN SPACE   113

If B^i_{jkl} is contracted with respect to the indices i and l, the resulting tensor is called the *Ricci Tensor* and is denoted by R_{jk}. Thus

$$R_{jk} = B^i_{jki}. \tag{37.23}$$

This tensor has an important role to play in Einstein's theory of gravitation. Since B^i_{jkl} is skew-symmetric with respect to the indices k and l, its contraction with respect to i and k yields only the Ricci tensor again in the form $-R_{jl}$. However, contraction with respect to the indices i and j yields another second rank tensor, viz.

$$S_{kl} = B^i_{ikl} = \frac{\partial \Gamma^i_{il}}{\partial x^k} - \frac{\partial \Gamma^i_{ik}}{\partial x^l}. \tag{37.24}$$

38. Geodesic coordinates. The Bianchi identities

If the affinity Γ^i_{jk} is symmetric in its subscripts, it is always possible to find a coordinate frame in which all components of the affinity vanish at any chosen point.

We shall first choose a coordinate frame in which the chosen point P is the origin and hence has coordinates $x^i = 0$. We then transform to new coordinates $\bar{x}^i$ by the equations

$$x^i = \bar{x}^i + \tfrac{1}{2} a^i_{jk} \bar{x}^j \bar{x}^k, \tag{38.1}$$

where the a^i_{jk} are constants whose values will be chosen later and we shall assume, without loss of generality, that a^i_{jk} is symmetric with respect to j and k. Differentiating the equations (38.1), we find that

$$\left. \begin{array}{l} \dfrac{\partial x^i}{\partial \bar{x}^j} = \delta^i_j + a^i_{jk} \bar{x}^k, \\[6pt] \dfrac{\partial^2 x^i}{\partial \bar{x}^j \partial \bar{x}^k} = a^i_{jk}. \end{array} \right\} \tag{38.2}$$

At P $x^i = 0$ and these reduce to

$$\frac{\partial x^i}{\partial \bar{x}^j} = \delta^i_j, \quad \frac{\partial^2 x^i}{\partial \bar{x}^j \partial \bar{x}^k} = a^i_{jk}. \tag{38.3}$$

Now
$$\frac{\partial x^i}{\partial \bar{x}^j} \frac{\partial \bar{x}^j}{\partial x^k} = \delta^i_k \tag{38.4}$$

and hence, at the chosen point,

$$\delta^i_j \frac{\partial \bar{x}^j}{\partial x^k} = \delta^i_k, \tag{38.5}$$

or
$$\frac{\partial \bar{x}^i}{\partial x^k} = \delta^i_k. \tag{38.6}$$

Substituting appropriately in equation (34.8), we calculate that the components of the affinity at the point P in the $\bar{x}^i$-frame are given by

$$\begin{aligned}\bar{\Gamma}^l_{ij} &= \delta^l_r \delta^s_i \delta^t_j \Gamma^r_{st} + \delta^l_r a^r_{ij}, \\ &= \Gamma^l_{ij} + a^l_{ij}. \end{aligned} \tag{38.7}$$

Since the affinity is symmetric, it is now possible to choose the a^l_{ij} to satisfy

$$a^l_{ij} = -\Gamma^l_{ij}. \tag{38.8}$$

The transformation (38.1) is now determined completely and, by equation (38.7),

$$\bar{\Gamma}^l_{ij} = 0 \tag{38.9}$$

as required.

The coordinates $\bar{x}^i$ are said to be *geodesic* at the point P. If such coordinates are employed, it is clear that covariant and partial derivatives are identical at P. This enables us to simplify many arguments leading to tensor equations. However, if such equations can, by this means, be proved valid in the geodesic frame, they are necessarily valid in all frames. As an example, we will derive the *Bianchi Identity*.

Thus, assuming the affinity is symmetric and employing geodesic coordinates at the point being considered, the covariant derivative of equation (37.21) is simply

$$\begin{aligned}B^i_{jkl;m} &= \frac{\partial}{\partial x^m}\left(\Gamma^i_{rk}\Gamma^r_{jl} - \Gamma^i_{rl}\Gamma^r_{jk} + \frac{\partial \Gamma^i_{jl}}{\partial x^k} - \frac{\partial \Gamma^i_{jk}}{\partial x^l}\right), \\ &= \frac{\partial^2 \Gamma^i_{jl}}{\partial x^m \partial x^k} - \frac{\partial^2 \Gamma^i_{jk}}{\partial x^m \partial x^l}, \end{aligned} \tag{38.10}$$

since the Γ^i_{jk} (but not their derivatives necessarily) all vanish at this point. Cyclically permuting the indices k, l, m in equation (38.10), we obtain

$$B^i_{jlm;k} = \frac{\partial^2 \Gamma^i_{jm}}{\partial x^k \partial x^l} - \frac{\partial^2 \Gamma^i_{jl}}{\partial x^k \partial x^m}, \tag{38.11}$$

$$B^i_{jmk;l} = \frac{\partial^2 \Gamma^i_{jk}}{\partial x^l \partial x^m} - \frac{\partial^2 \Gamma^i_{jm}}{\partial x^l \partial x^k}. \tag{38.12}$$

Addition of equations (38.10), (38.11), (38.12), yields the following identity

$$B^i_{jkl;m} + B^i_{jlm;k} + B^i_{jmk;l} = 0. \tag{38.13}$$

But this is a tensor equation and, having been proved true in the geodesic frame, must be true in all frames. Also, since the chosen point can be any point of $\mathscr{S}_N$, it is valid at all points of the space. It is the Bianchi Identity.

39. Metrical connection. Raising and lowering indices

In this section we shall further particularize our space $\mathscr{S}_N$ by supposing it to be Riemannian. That is, we shall suppose that a 'distance' or *interval ds* between two neighbouring points x^i, $x^i + dx^i$ is defined by the equation

$$ds^2 = g_{ij} dx^i dx^j, \tag{39.1}$$

where the N^2 coefficients g_{ij} are specified in some coordinate frame at every point of $\mathscr{S}_N$. It will be assumed, without loss of generality, that the g_{ij} are symmetric. Such a relationship between all pairs of adjacent points is called a *metrical connection* and the expression (39.1) for ds^2 is termed the *metric*.

For any two neighbouring points, ds will be regarded as an invariant associated with them and the g_{ij} must accordingly transform so that this shall be so. Since $dx^i dx^j$ is an arbitrary symmetric tensor, g_{ij} is symmetric and ds^2 is an invariant, it follows by a modified quotient theorem similar to the one proved in section 37, that g_{ij} is a tensor. It is called the *fundamental covariant tensor*. The contravariant tensor which is conjugate to g_{ij} (see section 31), viz. g^{ij}, is termed the *fundamental contravariant tensor*. This exists only if $g = |g_{ij}| \neq 0$, which we accordingly assume to be the case.

Let A^i be any contravariant vector defined at a point of $\mathscr{R}_N$. Then $g_{ij}A^j$ is a covariant vector at the same point and this will be denoted by A_i. Thus

$$A_i = g_{ij}A^j. \tag{39.2}$$

We shall regard the A^i and A_i as the contravariant and covariant components respectively, relative to the coordinate frame in use, of the *same* vector. The process defined by equation (39.2), of converting the contravariant expression for a vector into its covariant expression is termed *lowering the index*.

If B_i is a covariant vector, its contravariant expression is determined by *raising the index* with the aid of the fundamental contravariant vector. Thus

$$B^i = g^{ij}B_j. \tag{39.3}$$

For the notation to be consistent, it is necessary that if an index is first lowered and then raised, the original vector should again be obtained. This is seen to be the case for, if A_i is formed from A^i (equation (39.2)), the result of raising its subscript is (equation (39.3))

$$g^{ij}A_j = g^{ij}g_{jk}A^k = \delta^i_k A^k = A^i, \tag{39.4}$$

where equations (31.9) have been used in the reduction. Similarly, if an index is first raised and then lowered, the original covariant vector is reproduced.

Any index of a tensor can now be raised or lowered in the obvious way. Thus, if $A^{ij}{}_k$ is a tensor, we define

$$A^i{}_{jk} = g_{jr}A^{ir}{}_k. \tag{39.5}$$

To allow for the possibility that indices may be raised or lowered during a calculation, it will be convenient to displace the subscripts to the right of the superscripts. It is also often helpful to keep a record of these operations by placing a dot in the gap resulting from the raising or lowering of an index. These conventions are illustrated in equation (39.5).

Suppose an index of the fundamental tensor g_{ij} is raised. The result is

$$g^k{}_{.j} = g^{ki}g_{ij} = \delta^k{}_j, \tag{39.6}$$

i.e. the mixed fundamental tensor. The same tensor results when an

GENERAL TENSOR CALCULUS. RIEMANNIAN SPACE 117

index of g^{ij} is lowered. If both subscripts of g_{ij} are raised, the result is

$$g^{ri}g^{sj}g_{ij} = g^{ri}\delta^s{}_i = g^{rs}. \tag{39.7}$$

Our notation is entirely consistent, therefore, and g_{ij}, g^{ij}, $\delta^i{}_j$ are taken to be the covariant, contravariant and mixed components respectively of a single fundamental tensor.

Consider the inner product of two vectors A^i, B_i. We have

$$\begin{aligned} A^i B_i &= g^{ij} A_j g_{ik} B^k, \\ &= g^{ij} g_{ik} A_j B^k, \\ &= \delta^j{}_k A_j B^k, \\ &= A_k B^k, \\ &= A_i B^i. \end{aligned} \tag{39.8}$$

It is clear that the dummy index occurring in the expression for an inner product can be raised in one factor and lowered in the other without affecting the result. This is obviously valid for the inner product of any pair of tensors.

In $\mathscr{E}_N$, if we confine ourselves to rectangular Cartesian frames,

$$ds^2 = dx^i dx^i \tag{39.9}$$

and hence

$$\left.\begin{aligned} g_{ij} &= 1, \quad i = j, \\ &= 0, \quad i \neq j. \end{aligned}\right\} \tag{39.10}$$

It will now be found that $g = |g_{ij}| = 1$ and that

$$\left.\begin{aligned} g^{ij} &= 1, \quad i = j. \\ &= 0, \quad i \neq j. \end{aligned}\right\} \tag{39.11}$$

Hence, if A_i is a vector in the space, according to our definition its contravariant components will be

$$A^i = g^{ij} A_j = A_i, \tag{39.12}$$

confirming that there is no distinction between covariant and contravariant vectors in this special case of $\mathscr{E}_N$.

40. Scalar products. Magnitudes of vectors

In $\mathscr{E}_N$, the magnitude of the displacement vector dx^i is taken to be ds as given by equation (39.9). In $\mathscr{R}_N$, the magnitude of this vector is taken to be ds as given by equation (39.1). If A^i is any other contravariant vector, it may be represented as a displacement vector and then its *magnitude* is the invariant A, where

$$A^2 = g_{ij} A^i A^j. \tag{40.1}$$

This equation is accordingly taken to define the magnitude of A^i.

Raising and lowering the dummy indices in equation (40.1), we obtain the equivalent result

$$A^2 = g^{ij} A_i A_j. \tag{40.2}$$

It is natural to assume that the associated vectors A_i, A^i have equal magnitudes and hence A is also taken to be the magnitude of A_i. Equation (40.2) indicates how this can be calculated directly from A_i.

Since $g_{ij} A^j = A_i$ and $g^{ij} A_j = A^i$, equations (40.1), (40.2) are also both seen to be equivalent to the equation

$$A^2 = A_i A^i. \tag{40.3}$$

The *scalar product* of two vectors $\mathbf{A}$, $\mathbf{B}$ is defined to be the invariant

$$\mathbf{A} \cdot \mathbf{B} = A_i B^i = A^i B_i = g_{ij} A^i B^j = g^{ij} A_i B_j. \tag{40.4}$$

It will be noted that
$$A^2 = \mathbf{A} \cdot \mathbf{A}. \tag{40.5}$$

By analogy with $\mathscr{E}_3$, we now define the angle θ between two vectors $\mathbf{A}$, $\mathbf{B}$ to be such that

$$AB \cos \theta = \mathbf{A} \cdot \mathbf{B}. \tag{40.6}$$

I.e.
$$\cos \theta = \frac{A^i B_i}{\sqrt{[(A^j A_j)(B^k B_k)]}}. \tag{40.7}$$

If $\theta = \tfrac{1}{2}\pi$, the vectors are said to be *orthogonal* and

$$A^i B_i = 0. \tag{40.8}$$

41. The Christoffel symbols. Metric affinity

At any chosen point P of $\mathscr{R}_N$, as is proved in algebra texts, the homogeneous quadratic form defining the metric can, by a regular linear

GENERAL TENSOR CALCULUS. RIEMANNIAN SPACE 119

transformation from the coordinates x^i to coordinates y^i, be reduced to diagonal form thus

$$g_{ij}dx^i dx^j = (dy^1)^2 + (dy^2)^2 + \ldots + (dy^N)^2. \tag{41.1}$$

The transformation may involve complex coefficients. The metric may be assumed to take this simplified form over a small neighbourhood of P in which the y^i will behave like rectangular Cartesian coordinates in $\mathscr{E}_N$. In such a neighbourhood, therefore, the components of the affinity Γ^l_{ik} will be representable in the form given by equation (33.8) and hence must be symmetric in their subscripts at the point P. This will be true at every point of $\mathscr{R}_N$. We have established, therefore, that an affinity imposed upon a Riemannian space must be symmetric if parallel displacement in any small Euclidean region is to be in accordance with the usual definition for such a region.

Apart from the fact that it is symmetric, the affinity is otherwise arbitrary. However, consider the covariant derivative $A_{i;k}$. Formally, this can be obtained from A^i in either of two ways, (i) by lowering the index i and then differentiating or (ii) by differentiating and then lowering the index. Unless these two operations commute, so that either process leads to the same result, confusion will clearly arise. It is convenient, therefore, to define the affinity, if possible, in such a way that these operations are commutative. Thus, we shall require that

$$g_{ij}A^j{}_{;k} = (g_{ij}A^j)_{;k}. \tag{41.2}$$

Expanding the right-hand member by differentiation of the factors of the product, the condition becomes

$$g_{ij}A^j{}_{;k} = g_{ij}A^j{}_{;k} + g_{ij;k}A^j,$$

or $$g_{ij;k}A^j = 0. \tag{41.3}$$

This is to be true for arbitrary A^j and thus

$$g_{ij;k} = 0. \tag{41.4}$$

I.e. the affinity should be chosen so that the covariant derivative of the metric tensor g_{ij} is identically zero. This can be done as follows:

Cyclically permuting the indices i, j and k in equation (41.4), we derive the equations

$$\left.\begin{array}{l} \dfrac{\partial g_{ij}}{\partial x^k} - \Gamma^r_{ik} g_{rj} - \Gamma^r_{jk} g_{ir} = 0, \\[2mm] \dfrac{\partial g_{jk}}{\partial x^i} - \Gamma^r_{ji} g_{rk} - \Gamma^r_{ki} g_{jr} = 0, \\[2mm] \dfrac{\partial g_{ki}}{\partial x^j} - \Gamma^r_{kj} g_{ri} - \Gamma^r_{ij} g_{kr} = 0. \end{array}\right\} \quad (41.5)$$

Adding the last pair of this set of equations and subtracting the first, and remembering that g_{ij}, Γ^i_{jk} are symmetric, it will be found that

$$g_{kr} \Gamma^r_{ij} = [ij, k], \quad (41.6)$$

where
$$[ij, k] = \frac{1}{2}\left(\frac{\partial g_{jk}}{\partial x^i} + \frac{\partial g_{ki}}{\partial x^j} - \frac{\partial g_{ij}}{\partial x^k}\right). \quad (41.7)$$

$[ij, k]$ is termed the *Christoffel Symbol of the First Kind*. It is not a tensor, but its indices invariably behave like subscripts in any formula in which it occurs. It is symmetric in the indices i, j.

Multiplying both members of equation (41.6) by g^{sk} and summing with respect to k, it follows that

$$\Gamma^s_{ij} = \{^s_{ij}\}, \quad (41.8)$$

where $\quad \{^s_{ij}\} = g^{sk}[ij, k] = \tfrac{1}{2} g^{sk}\left(\dfrac{\partial g_{jk}}{\partial x^i} + \dfrac{\partial g_{ki}}{\partial x^j} - \dfrac{\partial g_{ij}}{\partial x^k}\right). \quad (41.9)$

$\{^s_{ij}\}$ is called *Christoffel's Symbol of the Second Kind*. It, also, is not a tensor and is symmetric with respect to the indices i, j.

It is now easy to verify that, if the affinity is determined by equation (41.8), the condition (41.4) is satisfied. Further, since

$$g^{ij} g_{kj} = \delta^i{}_k, \quad (41.10)$$

by taking the covariant derivative of both members of this equation, we obtain
$$g^{ij}{}_{;l} g_{kj} = 0. \quad (41.11)$$

GENERAL TENSOR CALCULUS. RIEMANNIAN SPACE 121

Multiplying by g^{kr} and summing with respect to k, we then find that

$$g^{ir}{}_{;l} = 0. \tag{41.12}$$

Thus, the covariant derivative of the fundamental contravariant tensor is also identically zero. It now follows that

$$(g^{ij}A_j)_{;k} = g^{ij}A_{j;k}, \tag{41.13}$$

i.e. a subscript may be raised before or after a covariant differentiation without affecting the result. This is clearly true for tensors of any rank.

The affinity determined by equation (41.8) will be called the *metric affinity*. With this choice of affinity, all forms of the fundamental tensor may be treated as constants in covariant differentiation.

In $\mathscr{E}_N$, if a rectangular Cartesian coordinate frame is employed, the g_{ij} are constants (equations (39.10)) and the Christoffel symbols are all identically zero. It follows, therefore, that covariant derivatives then reduce to ordinary partial derivatives. Such derivatives were proved to be tensors relative to these frames in section 11.

42. The covariant curvature tensor

The components of B^i_{jkl} are not all independent since the tensor is skew-symmetric in the indices k, l. In addition, however, *if the affinity is symmetric*, it is easily verified from equation (37.21) that

$$B^i_{jkl} + B^i_{klj} + B^i_{ljk} = 0. \tag{42.1}$$

If the affinity is metrical, by lowering the contravariant index of the Riemann-Christoffel tensor, a completely covariant curvature tensor B_{ijkl} is derived. This has a number of symmetry properties, one of which is obtained from our last equation immediately by lowering the index i throughout to give

$$B_{ijkl} + B_{iklj} + B_{iljk} = 0. \tag{42.2}$$

Further such properties can be established by first calculating a particular expression for this covariant tensor.

From equation (37.21), it follows that

$$B_{ijkl} = g_{is}\left[\{{}^s_r{}_k\}\{{}^r_j{}_l\} - \{{}^s_r{}_l\}\{{}^r_j{}_k\} + \frac{\partial}{\partial x^k}\{{}^s_j{}_l\} - \frac{\partial}{\partial x^l}\{{}^s_j{}_k\}\right]. \tag{42.3}$$

Multiplying equation (41.9) through by g_{rs} and summing with respect to s, we obtain the result

$$g_{rs}\{_i{}^s{}_j\} = \delta_r^k[ij,k] = [ij,r]. \qquad (42.4)$$

I.e. lowering an index of the symbol of the second kind, yields the symbol of the first kind. Equation (42.3) is therefore equivalent to

$$B_{ijkl} = [rk,i]\{_j{}^r{}_l\} - [rl,i]\{_j{}^r{}_k\} + \frac{\partial}{\partial x^k}[g_{is}\{_j{}^s{}_l\}] -$$
$$- \frac{\partial}{\partial x^l}[g_{is}\{_j{}^s{}_k\}] - \frac{\partial g_{is}}{\partial x^k}\{_j{}^s{}_l\} + \frac{\partial g_{is}}{\partial x^l}\{_j{}^s{}_k\}. \qquad (42.5)$$

Now, from equation (41.7), it follows that

$$\frac{\partial g_{is}}{\partial x^k} = [ik,s] + [sk,i] \qquad (42.6)$$

and hence

$$B_{ijkl} = [rk,i]\{_j{}^r{}_l\} - [rl,i]\{_j{}^r{}_k\} + \frac{\partial}{\partial x^k}[jl,i] - \frac{\partial}{\partial x^l}[jk,i] -$$
$$- \{_j{}^s{}_l\}([ik,s] + [sk,i]) + \{_j{}^s{}_k\}([il,s] + [sl,i]),$$
$$= \frac{\partial}{\partial x^k}[jl,i] - \frac{\partial}{\partial x^l}[jk,i] - \{_j{}^s{}_l\}[ik,s] + \{_j{}^s{}_k\}[il,s],$$
$$= \frac{1}{2}\left(\frac{\partial^2 g_{li}}{\partial x^j \partial x^k} + \frac{\partial^2 g_{jk}}{\partial x^i \partial x^l} - \frac{\partial^2 g_{jl}}{\partial x^i \partial x^k} - \frac{\partial^2 g_{ki}}{\partial x^j \partial x^l}\right) +$$
$$+ g_{sr}\{_i{}^r{}_l\}\{_j{}^s{}_k\} - g_{sr}\{_i{}^r{}_k\}\{_j{}^s{}_l\}. \qquad (42.7)$$

It is now clear that $\qquad B_{ijkl} = -B_{jikl}, \qquad (42.8)$

$$B_{ijkl} = -B_{ijlk}, \qquad (42.9)$$

$$B_{ijkl} = B_{klij}. \qquad (42.10)$$

Equations (42.8), (42.9) indicate that B_{ijkl} is skew-symmetric in its first and final pair of indices.

Also, lowering the superscript i throughout the Bianchi identity (38.13), we obtain

$$B_{ijkl;m} + B_{ijlm;k} + B_{ijmk;l} = 0. \qquad (42.11)$$

GENERAL TENSOR CALCULUS. RIEMANNIAN SPACE 123

43. Divergence. The Laplacian. Einstein's tensor

If the covariant derivative of a tensor field is found and then contracted with respect to the index of differentiation and any superscript, the result is called a *divergence* of the tensor. With respect to orthogonal coordinate transformations in $\mathscr{E}_N$, the partial and covariant derivatives are identical and then this definition of divergence agrees with that given in section 12.

From the tensor $A^{ij}{}_k$, two divergences can be formed, viz.

$$\operatorname{div}_i A^{ij}{}_k = A^{ij}{}_{k;i} \quad \text{and} \quad \operatorname{div}_j A^{ij}{}_k = A^{ij}{}_{k;j}. \tag{43.1}$$

A contravariant vector possesses one divergence only, which is an invariant. If the affinity is the metrical one, such a divergence is simply expressed in terms of ordinary partial derivatives thus: Since a derivative of a determinant can be found by differentiating each row separately and summing the results, we deduce that

$$\frac{\partial g}{\partial x^j} = G^{ik}\frac{\partial g_{ik}}{\partial x^j} = gg^{ik}\frac{\partial g_{ik}}{\partial x^j}. \tag{43.2}$$

Substituting for $\partial g_{ik}/\partial x^j$ from equation (42.6), this reduces to

$$\frac{\partial g}{\partial x^j} = gg^{ik}([ij,k]+[kj,i]) = 2g\{{}^i_{ij}\}. \tag{43.3}$$

Hence
$$\{{}^i_{ij}\} = \frac{1}{\sqrt{g}}\frac{\partial}{\partial x^j}(\sqrt{g}). \tag{43.4}$$

Now let A^i be a vector field. Its divergence is

$$\begin{aligned} A^i{}_{;i} &= \frac{\partial A^i}{\partial x^i} + \{{}^i_{ji}\}A^j, \\ &= \frac{1}{\sqrt{g}}\left[\sqrt{g}\frac{\partial A^i}{\partial x^i} + A^j\frac{\partial}{\partial x^j}(\sqrt{g})\right], \\ &= \frac{1}{\sqrt{g}}\frac{\partial}{\partial x^i}(\sqrt{g}A^i), \end{aligned} \tag{43.5}$$

which is the expression required.

In particular, if the vector field is obtained from an invariant V by taking its gradient, we have

$$A_i = \frac{\partial V}{\partial x^i} \tag{43.6}$$

and hence
$$A^i = g^{ij}\frac{\partial V}{\partial x^j}. \tag{43.7}$$

From equation (43.5), it now follows that the divergence of this vector is

$$\operatorname{div}\operatorname{grad} V = \nabla^2 V = \frac{1}{\sqrt{g}}\frac{\partial}{\partial x^i}\left(\sqrt{g}\,g^{ij}\frac{\partial V}{\partial x^j}\right). \tag{43.8}$$

The right-hand member of this equation represents the form taken by the *Laplacian* of V in a general Riemannian space. In $\mathscr{E}_N$, employing rectangular axes, the equations (39.11) are valid and

$$\nabla^2 V = \frac{\partial^2 V}{\partial x^i \partial x^i}, \tag{43.9}$$

which is its familiar form.

We shall now calculate the divergence of the Ricci tensor R_{jk} (equation (37.23)).

If the metric affinity is being employed, this tensor is symmetric for

$$\begin{aligned}R_{kj} &= B^i{}_{kji} = g^{ir}B_{rkji} = g^{ir}B_{jirk} = g^{ir}B_{ijkr}\\ &= B^r{}_{jkr} = R_{jk},\end{aligned} \tag{43.10}$$

having employed equations (42.8)–(42.10). Raising either index accordingly yields the same mixed tensor R^j_k. If this is contracted, an invariant

$$R = R^j_j \tag{43.11}$$

is obtained. R is called the *curvature scalar* of $\mathscr{R}_N$.

The divergence of the Ricci tensor is

$$\begin{aligned}R^m_{l;m} &= g^{mi}R_{li;m},\\ &= g^{mi}B^j{}_{lij;m},\\ &= g^{mi}g^{jk}B_{klij;m}.\end{aligned} \tag{43.12}$$

In view of equation (42.10), the Bianchi identity (42.11) can be written in the form

$$B_{klij;m} + B_{lmij;k} + B_{mkij;l} = 0 \tag{43.13}$$

and it follows that

$$\begin{aligned}
R^m_{l;m} &= -g^{mi}g^{jk}(B_{lmij;k} + B_{mkij;l}), \\
&= -g^{mi}g^{jk}(B_{mlji;k} - B_{mkji;l}), \\
&= -g^{jk}(B^i{}_{lji;k} - B^i{}_{kji;l}), \\
&= -g^{jk}(R_{lj;k} - R_{kj;l}), \\
&= -R^k_{l;k} + R^j_{j;l}.
\end{aligned} \tag{43.14}$$

Thus
$$R^m_{l;m} = \tfrac{1}{2} R^j_{j;l} = \frac{1}{2}\frac{\partial R}{\partial x^l} \tag{43.15}$$

is the divergence of the Ricci tensor.

Consider now the mixed tensor

$$R^i_j - \tfrac{1}{2}\delta^i_j R. \tag{43.16}$$

Its divergence is
$$R^i_{j;i} - \tfrac{1}{2}\delta^i_j \frac{\partial R}{\partial x^i},$$

$$= R^i_{j;i} - \tfrac{1}{2}\frac{\partial R}{\partial x^j},$$

$$= 0. \tag{43.17}$$

This is *Einstein's Tensor*. Its covariant and contravariant components are

$$R_{ij} - \tfrac{1}{2}g_{ij}R, \quad R^{ij} - \tfrac{1}{2}g^{ij}R, \tag{43.18}$$

respectively. Upon contraction, it yields the invariant

$$R - \tfrac{1}{2}NR = -\tfrac{1}{2}(N-2)R. \tag{43.19}$$

44. Geodesics

Let C be any curve constructed in a space $\mathscr{R}_N$ having metric (39.1) and let s be a parameter defined on C such that, if $s, s+ds$ are its values at the respective neighbouring points P, P' on C, then ds is the interval between these two points. If x^i are the coordinates of any point P on C, then the curve will be defined by parametric equations

$$x^i = x^i(s). \tag{44.1}$$

Since dx^i are the components of a vector and ds is an invariant, dx^i/ds is a contravariant vector at P. Its magnitude is, by equation (40.1),

$$\left(g_{ij}\frac{dx^i}{ds}\frac{dx^j}{ds}\right)^{1/2} \tag{44.2}$$

and this is unity by equation (39.1). dx^i/ds is termed the *unit tangent* to the curve at P, its direction being that of the displacement dx^i along the curve from P.

Suppose C possesses the property that the tangents at all its points are parallel, i.e. the curve's direction is constant over its whole length. This property is clearly quite independent of the coordinate frame being employed. In $\mathscr{E}_3$, such a curve would, of course, be a straight line. In $\mathscr{R}_N$, the curve will be called a *geodesic*. A geodesic is accordingly the counterpart of the Euclidean straight line in a Riemannian space. Suppose P, P' are neighbouring points on a geodesic having coordinates x^i, $x^i + dx^i$ respectively. If the unit tangent at P is parallel displaced to P', it will then be identical with the actual unit tangent at this point. Now, by equation (35.4), after parallel displacement from P to P', the unit tangent has components

$$\frac{dx^i}{ds} + \delta\left(\frac{dx^i}{ds}\right) = \frac{dx^i}{ds} - \Gamma^i_{jk}\frac{dx^j}{ds}dx^k. \tag{44.3}$$

But the actual unit tangent for the point P' has components

$$\left(\frac{dx^i}{ds}\right)_{s+ds} = \frac{dx^i}{ds} + \frac{d^2x^i}{ds^2}ds. \tag{44.4}$$

The vectors (44.3), (44.4) are identical provided

$$\frac{d^2x^i}{ds^2} + \Gamma^i_{jk}\frac{dx^j}{ds}\frac{dx^k}{ds} = 0. \tag{44.5}$$

If these equations are satisfied at every point of the curve (44.1), it is a geodesic. We shall assume, in future, that the affinity is metrical.

GENERAL TENSOR CALCULUS. RIEMANNIAN SPACE 127

The N equations (44.5) are second order differential equations for the functions $x^i(s)$ and their solution will involve $2N$ arbitrary constants. If A, B are two given points having coordinates $x^i = a^i$, $x^i = b^i$ respectively, the $2N$ conditions that the geodesic must contain these points will, in general, determine the arbitrary constants. Hence there is, in general, a unique geodesic connecting every pair of points. However, in some cases, this will not be so. For example, the geodesics on the surface of a sphere ($\mathcal{R}_2$) are great circles and, in general, there are two great circle arcs joining two given points, a major arc and a minor arc. Also, if these points are diametrically opposed to one another, there is an infinity of great circle arcs connecting them.

Since dx^i/ds is everywhere a *unit* vector, on a geodesic

$$g_{ij}\frac{dx^i}{ds}\frac{dx^j}{ds} = 1. \tag{44.6}$$

This must, accordingly, be a first integral of the equations (44.5). To show that this is the case, multiply equations (44.5) through by $2g_{ir}dx^r/ds$ and sum with respect to i to obtain

$$2g_{ir}\frac{dx^r}{ds}\frac{d^2x^i}{ds^2} + 2g_{ir}\Gamma^i_{jk}\frac{dx^j}{ds}\frac{dx^k}{ds}\frac{dx^r}{ds} = 0. \tag{44.7}$$

Now $$2g_{ir}\frac{dx^r}{ds}\frac{d^2x^i}{ds^2} = \frac{d}{ds}\left(g_{ir}\frac{dx^i}{ds}\frac{dx^r}{ds}\right) - \frac{dg_{ir}}{ds}\frac{dx^i}{ds}\frac{dx_r}{ds}. \tag{44.8}$$

Also $$2g_{ir}\Gamma^i_{jk}\frac{dx^j}{ds}\frac{dx^k}{ds}\frac{dx^r}{ds} = 2[jk,r]\frac{dx^j}{ds}\frac{dx^k}{ds}\frac{dx^r}{ds},$$

$$= ([jk,r]+[rk,j])\frac{dx^j}{ds}\frac{dx^k}{ds}\frac{dx^r}{ds},$$

$$= \frac{\partial g_{jr}}{\partial x^k}\frac{dx^k}{ds}\frac{dx^j}{ds}\frac{dx^r}{ds},$$

$$= \frac{dg_{jr}}{ds}\frac{dx^j}{ds}\frac{dx^r}{ds}. \tag{44.9}$$

By addition of equations (44.8) and (44.9), it will be seen that equation (44.7) can be expressed in the form

$$\frac{d}{ds}\left(g_{ij}\frac{dx^i}{ds}\frac{dx^j}{ds}\right) = 0. \quad (44.10)$$

Upon integration, there results the first integral

$$g_{ij}\frac{dx^i}{ds}\frac{dx^j}{ds} = \text{constant}. \quad (44.11)$$

The constant of integration must, of course, be taken to be unity.

The definition of a geodesic which has been given at the beginning of this section cannot be applied to the class of curves for which the interval ds between adjacent points vanishes. For such a curve, the parametric representation (44.1) is not appropriate and a unit tangent cannot be defined. Instead, suppose that a (1-1) correspondence is set up between the points of the curve and the values of an invariant λ in some interval $\lambda_0 \leq \lambda \leq \lambda_1$, so that parametric equations for the curve can be written

$$x^i = x^i(\lambda). \quad (44.12)$$

It will be assumed that the derivatives $dx^i/d\lambda$ all exist at each point of the curve. These derivatives constitute a contravariant vector and this has zero magnitude for, since $ds = 0$ along the curve,

$$g_{ij}\frac{dx^i}{d\lambda}\frac{dx^j}{d\lambda} = 0. \quad (44.13)$$

This vector will be in the direction of the displacement vector along the curve dx^i and will be called a *zero tangent* to the curve. The curve will be termed a *null geodesic* if the zero tangents at all points of the curve are parallel. This implies that, when the zero tangent at P is parallel displaced to the adjacent point P', it must be parallel to the zero tangent at this latter point, and since the magnitudes of these two vectors at P' are the same, they will be taken to be identical. The condition for this to be so is found, as before, to be

$$\frac{d^2x^i}{d\lambda^2} + \Gamma^i_{jk}\frac{dx^j}{d\lambda}\frac{dx^k}{d\lambda} = 0. \quad (44.14)$$

GENERAL TENSOR CALCULUS. RIEMANNIAN SPACE 129

These are, therefore, the equations of the null geodesics. It may now be shown, by an argument similar to that culminating in equation (44.11), that a first integral of these equations is

$$g_{ij}\frac{dx^i}{d\lambda}\frac{dx^j}{d\lambda} = \text{constant}. \quad (44.15)$$

In this case the constant must be zero.

Equation (44.5) may be put in an alternative form which is more convenient for particular calculations, as follows: Multiply through by $2g_{ri}$ and sum with respect to i; the resulting equation is equivalent to

$$\frac{d}{ds}\left(2g_{ri}\frac{dx^i}{ds}\right) - 2\frac{dg_{ri}}{ds}\frac{dx^i}{ds} + 2g_{ri}\Gamma^i_{jk}\frac{dx^j}{ds}\frac{dx^k}{ds} = 0. \quad (44.16)$$

Now

$$2\frac{dg_{ri}}{ds}\frac{dx^i}{ds} = 2\frac{\partial g_{ri}}{\partial x^k}\frac{dx^k}{ds}\frac{dx^i}{ds}$$

$$= \left(\frac{\partial g_{rj}}{\partial x^k} + \frac{\partial g_{rk}}{\partial x^j}\right)\frac{dx^j}{ds}\frac{dx^k}{ds} \quad (44.17)$$

and

$$2g_{ri}\Gamma^i_{jk} = [jk, r] = \frac{\partial g_{rj}}{\partial x^k} + \frac{\partial g_{rk}}{\partial x^j} - \frac{\partial g_{jk}}{\partial x^r}. \quad (44.18)$$

Equation (44.16) accordingly reduces to

$$\frac{d}{ds}\left(2g_{ri}\frac{dx^i}{ds}\right) - \frac{\partial g_{jk}}{\partial x^r}\frac{dx^j}{ds}\frac{dx^k}{ds} = 0. \quad (44.19)$$

Exercises 5

1. A_{ij} is a covariant tensor. If $B_{ij} = A_{ji}$, prove that B_{ij} is a covariant tensor. Deduce that, if A_{ij} is symmetric (or skew-symmetric) in one frame, it is symmetric (or skew-symmetric) in all. [Hint: The equations $A_{ij} = A_{ji}$, $A_{ij} = -A_{ji}$ are tensor equations.]

2. (x, y, z) are rectangular Cartesian coordinates of a point P in $\mathscr{E}_3$ and (r, θ, ϕ) are the corresponding spherical polars related to the Cartesians by equations (29.3). $\mathbf{A}$ is a contravariant vector defined at P having components (A^x, A^y, A^z) in the Cartesian frame and components (A^r, A^θ, A^ϕ) in the spherical polar frame. Express the polar

components in terms of the Cartesian components. $O1$, $O2$, $O3$ are rectangular Cartesian axes such that P lies on $O1$ and $O3$ lies in the plane Oxy. If (A^1, A^2, A^3) are the components of **A** in this Cartesian frame, show that

$$A^1 = A^r, \quad A^2 = rA^\theta, \quad A^3 = r\sin\theta A^\phi.$$

[Note: Assume the Cartesian axes are right-handed.]

3. If A_i is a covariant vector, verify that $B_{ij} = A_{,j} - A_{j,i}$ transforms like a covariant tensor. (This is curl **A**.) If **A** is the gradient of a scalar, verify that its curl vanishes.

4. If A_{ij} is a skew-symmetric covariant tensor, verify that

$$B_{ijk} = A_{ij,k} + A_{jk,i} + A_{ki,j}$$

transforms as a tensor.

5. g_{ij} is the metric tensor of an $\mathscr{R}_3$. If $g = |g_{ij}|$, show that e^{ijk}, e_{ijk}, where

$$e^{ijk} = \mathfrak{e}^{ijk}/\sqrt{g}, \quad e_{ijk} = \sqrt{(g)}\,\mathfrak{e}_{ijk},$$

are tensors in the $\mathscr{R}_3$. A_{ij} is a skew-symmetric tensor in the space. Deduce that $\tfrac{1}{2} e^{ijk} A_{jk}$ is a contravariant vector whose components are $A_{23}/\sqrt{g}$, $A_{31}/\sqrt{g}$, $A_{12}/\sqrt{g}$. In particular, taking $A_{ij} = B_i C_j - B_j C_i$, show that the contravariant vector is the vector product of the covariant vectors B_i, C_i, if the space is Euclidean. (If the space is not Euclidean, this is taken to be the *definition* of the vector product of covariant vectors.)

6. A^{ij} is a skew-symmetric tensor in $\mathscr{R}_3$. Show that $\tfrac{1}{2} e_{ijk} A^{jk}$ is a covariant vector whose components are $A^{23}\sqrt{g}$, $A^{31}\sqrt{g}$, $A^{12}\sqrt{g}$. Hence, define the vector product of two contravariant vectors.

7. Show that

$$A_{i;j} - A_{j;i} = A_{i,j} - A_{j,i}$$

provided the affinity is symmetric.

8. Show that

$$A_{i;jk} - A_{i;kj} = B^r_{ijk} A_r + (\Gamma^r_{kj} - \Gamma^r_{jk}) A_{i;r}$$

and deduce that B^r_{ijk} is a tensor and that covariant differentiations are commutative in a space for which $B^r_{ijk} = 0$ and the affinity is symmetric. Obtain the corresponding result for a contravariant vector A.

9. A is defined at the point x^i and is parallel displaced around a

small contour enclosing the point. Prove that the increment in A^i resulting from one circuit is given by

$$\Delta A_i = -\tfrac{1}{2} B^l_{ijk} A_l \alpha^{jk},$$

where α^{jk} is defined by equation (37.10).

10. The parametric equations of a curve in $\mathscr{S}_N$ are

$$x^i = x^i(t);$$

t is an invariant parameter. A tensor A^i_j is defined over a region containing the curve. P, P' are neighbouring points t, $t+\Delta t$ on the curve and ΔA^i_j is defined to be the difference between the actual value of the tensor at P' and the value of the tensor at P after it has been parallel displaced to P'. Prove that

$$\frac{DA_j}{Dt} = \lim_{\Delta t \to 0} \frac{\Delta A_j}{\Delta t} = A^i_{j;k} \frac{dx^k}{dt}.$$

(DA^i_j/Dt is called the *intrinsic derivative* of the tensor along the curve.)

11. Verify that $\{^{\ i}_{j\ k}\}$, as defined by equation (41.9), transforms as an affinity.

12. If A_{ij} is symmetric, prove that $A_{ij;k}$ is symmetric in i and j.

13. Show that the number of the components of B^i_{jkl} which may be assigned values arbitrarily is, in general, $\tfrac{1}{2}N^3(N-1)$. If the affinity is symmetric, show that this number is $\tfrac{1}{3}N^2(N^2-1)$. [Hint: Use equation (42.1).]

14. Show that the number of the components of B_{ijkl} which may be assigned values arbitrarily is $N^2(N^2-1)/12$. [Hint: Use equations (42.2), (42.8), (42.9), (42.10).]

15. By differentiating the equation

$$g^{ij} g_{jk} = \delta^i_k$$

with respect to x^l, show that

$$\frac{\partial g^{im}}{\partial x^l} = -g^{mk} g^{ij} \frac{\partial g_{jk}}{\partial x^l}$$

and hence that

$$\frac{\partial g^{im}}{\partial x^l} + g^{ij}\{^{\ m}_{j\ l}\} + g^{mj}\{^{\ i}_{j\ l}\} = 0.$$

Deduce that $g^{ij}{}_{;k} = 0$.

16. If the affinity is the metric one, prove that

$$R_{jk} = B^i_{jki} = -\frac{\partial}{\partial x^i}\{^i_{jk}\} + \frac{\partial^2}{\partial x^j \partial x^k}\log\sqrt{g} +$$
$$+ \{^i_{rk}\}\{^r_{ji}\} - \{^r_{jk}\}\frac{\partial}{\partial x^r}\log\sqrt{g},$$

$$S_{kl} = B_{ikl} = 0.$$

[Hint: Employ equation (43.4).] Deduce that R_{jk} is symmetric.

17. If θ, ϕ are co-latitude and longitude respectively on the surface of a sphere of unit radius, obtain the metric

$$ds^2 = d\theta^2 + \sin^2\theta\, d\phi^2$$

for the surface. Show that the only non-vanishing three index symbols for this $\mathscr{R}_2$ are

$$\{^1_{22}\} = -\sin\theta\cos\theta, \quad \{^2_{12}\} = \{^2_{21}\} = \cot\theta.$$

Show also that the only non-vanishing components of B_{ijkl} are

$$B_{1212} = -B_{1221} = B_{2121} = -B_{2112} = \sin^2\theta$$

and that the components of the Ricci tensor are given by

$$R_{12} = R_{21} = 0, \quad R_{11} = -1, \quad R_{22} = -\sin^2\theta.$$

Prove that the curvature scalar is given by $R = -2$.

18. Employing equation (43.8), obtain expressions for $\nabla^2 V$ in cylindrical and spherical polars.

19. In a certain coordinate system

$$\Gamma^i_{jk} = \delta^i_j \frac{\partial \phi}{\partial x^k} + \delta^i_k \frac{\partial \psi}{\partial x^j},$$

where ϕ, ψ are functions of position. Prove that B^i_{jkl} is a function of ψ only. If $\psi = -\log(a_i x^i)$ prove that

$$R_{jk} = B^i_{jki} = 0.$$

(L.U.)

20. In the $\mathscr{R}_2$ whose metric is

$$ds^2 = \frac{dr^2 + r^2 d\theta^2}{r^2 - a^2} - \frac{r^2 dr^2}{(r^2 - a^2)^2}, \quad (r > a),$$

prove that the differential equation of the geodesics may be written

$$a^2\left(\frac{dr}{d\theta}\right)^2 + a^2 r^2 = k^2 r^4,$$

where k^2 is a constant such that $k^2 = 1$ if, and only if, the geodesic is null. By putting $r\,d\theta/dr = \tan\phi$, show that if the space is mapped on a Euclidean plane in which r, θ are taken as polar coordinates, the geodesics are mapped as straight lines, the null-geodesics being tangents to the circle $r = a$.

(L.U.)

21. A 2-space has metric

$$ds^2 = g_{11}(dx^1)^2 + g_{22}(dx^2)^2.$$

Prove that

$$\frac{1}{g}B_{1212} = -\frac{1}{2\sqrt{g}}\left\{\frac{\partial}{\partial x^1}\left(\frac{1}{\sqrt{g}}\frac{\partial g_{22}}{\partial x^1}\right) + \frac{\partial}{\partial x^2}\left(\frac{1}{\sqrt{g}}\frac{\partial g_{11}}{\partial x^2}\right)\right\}.$$

(L.U.)

22. Prove that

(i) $$A^{ij}{}_{;i} = \frac{1}{\sqrt{g}}\frac{\partial}{\partial x^i}(\sqrt{g}A^{ij}) + A^{ik}\{{}^{\ j}_{i\ k}\}$$

(ii) $$X^{ij}{}_{;ij} = 0,$$

provided X^{ij} is skew-symmetric. Hence prove that, for any tensor A^{ij}

$$A^{ij}{}_{;ij} = A^{ij}{}_{;ji}.$$

(L.U.)

23. A curve C has parametric equations

$$x^i = x^i(t)$$

and joins two points A and B. The length of the curve is defined to be

$$L = \int_A^B ds = \int_A^B \sqrt{\left(g_{ij}\frac{dx^i}{dt}\frac{dx^j}{dt}\right)}\,dt.$$

Write down the Euler conditions that L should be stationary with respect to all small variations from C and by changing the independent variable in these conditions from t to s, show that they are identical with equations (44.5). (This provides an alternative definition for a geodesic.)

24. If Γ^i_{jk} is a symmetric affinity, show that

$$\Gamma^{i*}_{jk} = \Gamma^i_{jk} + \delta^i_j A_k + \delta^i_k A_j$$

is also a symmetric affinity.

If B^i_{jkl}, B^{i*}_{jkl} are the Riemann-Christoffel curvature tensors relative to the affinities Γ^i_{jk}, Γ^{i*}_{jk} respectively, prove that

$$B^{i*}_{jkl} = B^i_{jkl} + \delta^i_k A_{jl} - \delta^i_l A_{jk} + \delta^i_j(A_{kl} - A_{lk}),$$

where $A_{ij} = A_i A_j - A_{i;j}$.

Hence show that if A_i is the gradient of a scalar, then

$$B^{i*}_{iil} - B^{i*}_{lij} = B^i_{jil} - B^i_{lij}.$$

(M.T.)

25. Prove that the affinity transformations form a group.

26. If $D = |\partial x^i / \partial \bar{x}^j|$, show that

$$\frac{1}{D} \frac{\partial D}{\partial \bar{x}^i} = \frac{\partial \bar{x}^j}{\partial x^k} \frac{\partial^2 x^k}{\partial \bar{x}^i \partial \bar{x}^j}.$$

27. Show that the transformation law for the quantities K_i (equation (36.5)) is

$$\bar{K}_i = \frac{\partial x^j}{\partial \bar{x}^i} K_j + \frac{\partial \bar{x}^j}{\partial x^k} \frac{\partial^2 x^k}{\partial \bar{x}^i \partial \bar{x}^j}$$

and deduce that $\Gamma^r_{ri} - K_i$ is a tensor.

28. Oblique cartesian axes are taken in a plane. Show that the contravariant components of a vector **A** can be obtained by projecting a certain displacement vector on to the axes by parallels to the axes and the covariant components by projecting by perpendiculars to the axes.

29. Define coordinates (r, ϕ) on a right circular cone having semi-vertical angle α so that the metric for the surface is

$$ds^2 = dr^2 + r^2 \sin^2 \alpha \, d\phi^2.$$

Show that the family of geodesics is given by

$$r = a \sec(\phi \sin \alpha - \beta),$$

where a, β are arbitrary constants. Explain this result by developing the cone into a plane.

30. An $\mathscr{R}_N$ has metric

$$ds^2 = e^\lambda dx^i dx^i,$$

where λ is a function of the x^i. Show that the only non-vanishing Christoffel symbols of the second kind are

$$\begin{Bmatrix} P \\ QQ \end{Bmatrix} = (\delta_Q^P - \tfrac{1}{2}) \lambda_P, \quad \begin{Bmatrix} P \\ PQ \end{Bmatrix} = \tfrac{1}{2} \lambda_Q,$$

where $\lambda_r = \partial \lambda / \partial x^r$. Deduce that

$$\begin{Bmatrix} i \\ rP \end{Bmatrix} \begin{Bmatrix} r \\ Pi \end{Bmatrix} = \tfrac{1}{4}(N+2) \lambda_P^2 - \tfrac{1}{2} \lambda_r \lambda_r$$

and that the scalar curvature of this space is given by

$$R = (N-1) e^{-\lambda} [\lambda_{rr} + \tfrac{1}{4}(N-2) \lambda_r \lambda_r],$$

where $\lambda_{rr} = \partial^2 \lambda / \partial x^r \partial x^r$.

31. θ is the colatitude and ϕ is the longitude on a unit sphere, so that the metric for the surface is

$$ds^2 = d\theta^2 + \sin^2\theta \, d\phi^2.$$

The covariant vector A_i is taken with initial components (X, Y) and is carried, by parallel displacement, along an arc of length $\phi \sin \alpha$ of the circle $\theta = \alpha$. Show that the components of A_i attain the final values

$$A_1 = X \cos(\phi \cos \alpha) + Y \csc \alpha \sin(\phi \cos \alpha),$$
$$A_2 = -X \sin \alpha \sin(\phi \cos \alpha) + Y \cos(\phi \cos \alpha).$$

Verify that the magnitude of the vector A_i is unaltered by the displacement.

32. An $\mathscr{R}_3$ has metric

$$ds^2 = \lambda dr^2 + r^2(d\theta^2 + \sin^2\theta \, d\phi^2),$$

where λ is a function of r alone. Show that, along the geodesic for which $\theta = \tfrac{1}{2}\pi$, $d\theta/ds = 0$ at $s = 0$,

$$\phi = \int \lambda^{1/2} d\psi,$$

where $r = b \sec \psi$. Interpret this result geometrically when $\lambda = 1$.

33. y^i ($i = 1, 2, 3, 4$) are rectangular cartesian coordinates in $\mathscr{E}_4$. Show that

$$y^1 = R\cos\theta,$$
$$y^2 = R\sin\theta\cos\phi,$$
$$y^3 = R\sin\theta\sin\phi\cos\psi,$$
$$y^4 = R\sin\theta\sin\phi\sin\psi,$$

are parametric equations of a hypersphere of radius R. If (θ, ϕ, ψ) are taken as coordinates on the hypersphere, show that the metric for this $\mathscr{R}_3$ is

$$ds^2 = R^2[d\theta^2 + \sin^2\theta(d\phi^2 + \sin^2\phi\, d\psi^2)].$$

Deduce that in this $\mathscr{R}_3$,

$$B_{1212} = R^2\sin^2\theta, \quad B_{2323} = R^2\sin^4\theta\sin^2\phi,$$
$$B_{3131} = R^2\sin^2\theta\sin^2\phi,$$

all other distinct components being zero. Hence show that

$$B_{ijkl} = K(g_{ik}g_{jl} - g_{il}g_{jk})$$

where $K = 1/R^2$. (This is the condition for the space to be of constant Riemannian curvature K.)

34. An $\mathscr{R}_2$ has metric

$$ds^2 = \text{sech}^2 y(dx^2 + dy^2).$$

Find the equation of the family of geodesics.

θ, ϕ are colatitude and longitude respectively on the surface of a sphere of unit radius. Mercator's projection is obtained by plotting x, y as rectangular cartesian coordinates in a plane, taking

$$x = \phi, \quad y = \log\cot\tfrac{1}{2}\theta.$$

Calculate the metric for the spherical surface in terms of x and y and deduce that the great circles are represented by the curves

$$\sinh y = \alpha\sin(x + \beta),$$

where α, β are parameters, in Mercator's projection.

CHAPTER 6

General Theory of Relativity

45. Principle of equivalence

The special theory of relativity rejects the Newtonian concept of a privileged observer, at rest in absolute space, and for whom physical laws assume their simplest form and assumes, instead, that these laws will be identical for all members of a class of inertial observers in uniform translatory motion relative to one another. Thus, although the existence of a *single* privileged observer is denied, the existence of a *class* of such observers is accepted. This seems to imply that, if all matter in the universe were annihilated except for a single experimenter and his laboratory, this observer would, nonetheless, be able to distinguish inertial frames from non-inertial frames by the special simplicity which the descriptions of physical phenomena take with respect to the former. The further implication is, therefore, that physical space is not simply a mathematical abstraction which it is convenient to employ when considering distance relationships between material bodies, but exists in its own right as a separate entity with sufficient internal structure to permit the definition of inertial frames. However, all the available evidence suggests that physical space cannot be defined except in terms of distance measurements between physical bodies. For example, such a space can be constructed by setting up a rectangular Cartesian coordinate frame comprising three mutually perpendicular rigid rods and then defining the coordinates of the point occupied by a material particle by distance measurements from these rods in the usual way. Physical space is, then, nothing more than the aggregate of all possible coordinate frames. A claim that physical space exists independently of distance measurements between material bodies, can only be substantiated if a precise statement is given of the manner in which its existence can be detected without carrying out such measurements. This has never been done and we shall assume, therefore, that the special properties possessed by inertial frames must be related in some way to the distribution of matter within the universe and that they are not an indication of an inherent structure possessed by physical space when it is considered apart from the matter it contains. This line of argument encourages us to expect, therefore, that, ultimately, all physical laws will be expressible in forms which are quite independent of any co-ordinate frame by which physical space is defined, i.e. that physical laws are identical for all observers. This is the *general principle of relativity*. This does not mean that, when account is taken of the actual distribution of matter within the universe, certain frames will not prove to be more convenient than others. When calculating the field due to a distribution of electric charge, it simplifies the calculations enormously if a reference frame can be employed relative to which the charge is wholly at rest. However, this does not mean that the laws of electromagnetism are expressible more simply in this frame, but only that this particular charge distribution is then described more simply. Similarly, we shall attribute the simpler forms taken by some calculations when carried out in inertial frames, to the special relationship these frames bear to the matter present in the universe. Fundamentally, therefore, all observers will be regarded as equivalent and, by employing the same physical laws, will arrive at identical conclusions concerning the development of any physical system.

The main difficulty which arises when we try to express physical laws so that they are valid for all observers is that, if test particles are released and their motions studied from a frame which is being accelerated with respect to an inertial frame, these motions will not be uniform and this fact appears to set such frames apart from inertial frames as a special class for which the ordinary laws of motion do not apply. However, by a well-known device of Newtonian mechanics, viz. the introduction of *inertial forces*, accelerated frames can be treated as though they were inertial and this suggests a way out of our difficulty. Thus suppose a space rocket, moving *in vacuo*, is being accelerated uniformly by the action of its motors. An observer inside the rocket will note that unsupported particles experience an acceleration parallel to the axis of the rocket. Knowing that the motors are operating, he will attribute this acceleration to the fact that his natural reference frame is being accelerated relative to an inertial frame. However he may, if he prefers, treat his reference frame as inertial and suppose that all bodies within the rocket are being subjected to inertial forces acting parallel to the rocket's axis. If **f** is the acceleration of the rocket, the appropriate inertial force to be applied to a particle of mass m is $-m\mathbf{f}$. Similarly, if the rocket's motors are shut down but the rocket is spinning about its axis, an observer within the rocket will again note that free particles do not move uniformly relative to his surroundings and he may again avoid attributing this phenomenon to the fact that his frame is not inertial, by supposing certain inertial forces (viz. centrifugal and Coriolis forces) to act upon the particles. Now it is an obvious property of each such inertial force that it must cause an acceleration which is independent of the mass of the body upon which it acts, for the force is always obtained by multiplying the body's mass by an acceleration independent of the mass. This property it shares with a gravitational force, for this also is proportional to the mass of the particle being attracted and hence induces an acceleration which is independent of this mass. This independence of the gravitational acceleration of a particle and its mass has been checked experimentally with great accuracy by Eötvös. If, therefore, we regard the equivalence of inertial and gravitational forces as having been established, inertial forces can be thought of as arising from the presence of gravitational fields. This is the *Principle of Equivalence*. By this principle, in the case of the uniformly accelerated rocket, the observer is entitled to neglect his acceleration relative to an inertial frame, provided he accepts the existence of a uniform gravitational field of intensity $-\mathbf{f}$ parallel to the axis of the rocket. Similarly, the observer in the rotating rocket may disregard his motion and accept, instead, the existence of a gravitational field having such a nature as to account for the centrifugal and Coriolis forces.

By appeal to the principle of equivalence, therefore, an observer employing a reference frame in arbitrary motion with respect to an inertial frame, may disregard this motion and assume, instead, the existence of a gravitational field. The intensity of this field at any point within the frame will be equal to the inertial force per unit mass at the point. By this device, every observer becomes entitled to treat his reference frame as being at rest and all observers accordingly become equivalent. However, the reader is probably still not convinced that the distinction between accelerated and inertial frames has been effectively eliminated, but only that it has been concealed by means of a mathematical device having no physical significance. Thus, he may point out that the gravitational fields which have been introduced to account for the inertial forces are 'fictitious' fields, which may be completely removed by choosing an inertial frame for reference purposes, whereas 'real' fields, such as those due to the Earth and Sun, cannot be so removed. He may further object that no physical agency can be held responsible for the presence of a 'fictitious' field, whereas a 'real' field is caused by the presence of a massive body. These objections may be met by attributing such 'fictitious' fields to the motions of distant masses within the universe. Thus, if an observer within the uniformly accelerated rocket takes himself to be at rest, he must accept as an observable fact that all bodies within the universe, including the galaxies, possess an additional acceleration of $-\mathbf{f}$ relative to him and to this motion he will be able to attribute the presence of the uniform gravitational field which is affecting his test particles. Again, the whole universe will be in rotation about the observer who regards himself and his space-ship as stationary when it is in rotation relative to an inertial frame. It is this rotation of the masses of the universe which we shall hold responsible for the Coriolis and centrifugal gravitational fields within the rocket. But, in addition, these 'inertial' gravitational fields will account for the motions of the galaxies as observed from the noninertial frame. Thus, for the observer within the uniformly accelerated rocket a uniform gravitational field of intensity $-\mathbf{f}$ extends over the whole of space and is the cause of the acceleration of the galaxies; for the observer within the rotating rocket, the resultant of the centrifugal and Coriolis fields acting upon the galaxies is just sufficient to account for their accelerations in their circular orbits about himself as centre (the reader should verify this, employing the results of Exercise 1, Chapter 1). On this view, therefore, inertial frames possess particularly simple properties only because of their special relationship to the distribution of mass within the universe. In much the same way, the electromagnetic field due to a distribution of electric charge takes an especially simple form when described relative to a frame in which all the charges are at rest (assuming such exists). If any other frame is employed, the field will be complicated by the presence of a magnetic

GENERAL THEORY OF RELATIVITY 141

component arising from the motions of the charges. However, this magnetic field is not considered imaginary because a frame can be found in which it vanishes, whereas for certain magnetic fields such a frame cannot be found. The laws of electromagnetism are taken to be valid in all frames, though it is conceded that, for solving particular problems, a certain frame may prove to be pre-eminently more convenient than any other. Neither, therefore, should the centrifugal and Coriolis fields be dismissed as imaginary solely because they can be removed by proper choice of a reference frame, although it may be convenient to make such a choice of frame when carrying out particular computations. In short, the general principle of relativity can be accepted as valid and, at the same time, the existence of the inertial frames accounted for by the simplicity of the motions of the galactic masses with respect to these particular frames.

If it is accepted that the existence of inertial frames is bound up with the large-scale distribution of matter within the universe, then it follows that the inertia possessed by a body, which causes it to move uniformly in such a frame, is also a consequence of this matter distribution. This is *Mach's Principle*, viz. that mass is induced in a body by the presence of distant matter in the universe.

The previously unexplained identity of inertial and gravitational masses is easily deduced as a consequence of the principle of equivalence. For, consider a particle of mass m which is being observed from a non-inertial frame. A gravitational force equal to the inertial force will be observed to act upon this body. This force is directly proportional to the inertial mass m. But, by the principle of equivalence, all gravitational forces are of the same nature as this particular force and will, accordingly, be directly proportional to the inertial masses of the bodies upon which they act. Thus the gravitational 'charge' of a particle, measuring its susceptibility to the influences of gravitational fields, is identical with its inertial mass and the identity of inertial and gravitational masses has been explained in a straightforward and convincing manner.

46. Metric in a gravitational field

Suppose that an inertial frame has been established in a region of space where there are no local gravitational influences and that a material plane is set rotating with angular velocity ω relative to the frame about an axis perpendicular to it and fixed in the inertial frame. An observer O, moving with the plane, is entitled to consider himself as at rest in the presence of a gravitational field which will account for the centrifugal and Coriolis forces. Suppose O identifies a large number of points on the plane which he measures with a standard rod to be at the same distance r from the axis of rotation and, by this means, constructs a circle. If, now, O lays his measuring rod repeatedly along a radius of this circle to measure its length r and this operation is watched by an observer O' who is stationary in the inertial frame, O' will agree with the length found by O, for the rod will only move laterally during the process of measurement and hence will suffer no Fitzgerald contraction of its length. O' will therefore agree with O that all the radii are of equal length and hence that the figure constructed is a circle. However, if O now lays his rod along the circumference of the circle (we assume the dimensions of the rod are small by comparison with those of the circle) and measures its length to be l, O' will disagree with this measurement for the reason that, for him, throughout the measuring process the rod will have a speed ωr along its length and hence will have contracted by the factor $(1 - \omega^2 r^2/c^2)^{1/2}$. Allowing for this contraction, O' will assert the length of the circumference to be

$$l(1 - \omega^2 r^2/c^2)^{1/2}. \tag{46.1}$$

But O' is employing an inertial frame in which the geometry is known to be Euclidean. It follows that

$$l(1 - \omega^2 r^2/c^2)^{1/2} = 2\pi r \tag{46.2}$$

and hence that
$$l = \frac{2\pi r}{(1 - \omega^2 r^2/c^2)^{1/2}}. \tag{46.3}$$

This last equation implies that for O the ratio $l/2r$ is not π, but a number greater than π and the larger the radius of his constructed circle, the larger this ratio will become. By direct measurement, therefore, O will be able to ascertain that the geometry of figures drawn on the plane is not Euclidean. We conclude that, in the presence of a gravitational field of the centrifugal–Coriolis type, the geometry of space is not Euclidean.

GENERAL THEORY OF RELATIVITY 143

By the principle of equivalence, the conclusion which has just been reached concerning the non-Euclidean nature of space in which there is present a gravitational field of the centrifugal–Coriolis type, must be extended to all gravitational fields. However, in the case of a field such as that which surrounds the Earth, it will not be possible (as it is for the centrifugal–Coriolis field) to find an inertial frame of reference relative to which the field vanishes and for which the spatial geometry is Euclidean. Such a field will be termed *irreducible*. Even in an irreducible field, however, a frame can always be found which is inertial for a sufficiently small region of space and a sufficiently small time duration. Thus, within a space-ship which is not rotating relative to the extra-galactic nebulae and which is falling freely in the Earth's gravitational field, free particles will follow straight line paths at constant speed for considerable periods of time and the conditions will be inertial. A coordinate frame fixed in the ship will accordingly simulate an inertial frame over a restricted region of space and time and its geometry will be approximately Euclidean.

Since a rectangular Cartesian coordinate frame can be set up only in a space possessing a Euclidean metric, this method of specifying the relative positions of events must be abandoned in an irreducible gravitational field (except over small regions as has just been explained). Instead, it will be assumed that each physical event is allotted three space coordinates (ξ^1, ξ^2, ξ^3) according to any convenient scheme. It will be necessary in any particular case only to describe the procedure whereby these coordinates are to be established for any event. They might, for example, be determined by radar techniques. Thus radar transmitters could, in principle, be set up at three widely separated points and the electromagnetic pulses generated by them reflected back to their sources by the event in question. The time intervals between transmission of a pulse and reception of its 'echo' at the three stations, would then be suitable coordinates of the type we are considering for the event. All that is necessary is that there should exist a correspondence between points in space and triads of real numbers (ξ^1, ξ^2, ξ^3) such that, to each point there corresponds a unique triad, and to each triad there corresponds a unique point. Again, clocks will be supposed distributed throughout space so that to each event a time ξ^4 can be allocated, namely, the time indicated by the clock situated at the event. It will be assumed that, whatever system is adopted for fixing the coordinates ξ, a series of contiguous events, such as the positions occupied by a particle at successive instants of time, will have coordinates which will vary continuously along the series and which will correspond to a continuous curve in ξ-space. This implies that adjacent clocks must be synchronized and must run at the same not necessarily constant rate, but no special procedure for the synchronization of distant clocks need be laid down. It follows that, as in special relativity theory, two distant events may be simultaneous according to one system of time reckoning, but not so according to another.

We shall now further generalize the coordinates allocated to an event. Let x^i ($i = 1, 2, 3, 4$) be any functions of the ξ^i such that, to each set of values of the ξ^i there corresponds one set of values of the x, and conversely. We shall write

$$x^i = x^i(\xi^1, \xi^2, \xi^3, \xi^4). \tag{46.4}$$

Then the x^i, also, will be accepted as coordinates, with respect to a new frame of reference, of the event whose coordinates were previously taken to be the ξ. It should be noted that, in general, each of the new coordinates x^i will depend upon both the time and the position of the event, i.e., it will not necessarily be the case that three of the coordinates x^i are spatial in nature and one is temporal. All possible events will now be mapped upon a space $\mathscr{S}_4$, so that each event is represented by a point of the space and the x^i will be the coordinates of this point with respect to a coordinate frame. $\mathscr{S}_4$ will be referred to as the *space-time continuum*.

It has been remarked that, in any gravitational field, it is always possible to define a frame relative to which the field vanishes over a restricted region and which behaves as an inertial frame for events occurring in this region and extending over a small interval of time. Suppose, then, that such an inertial frame S is found for two contiguous events. Any other frame in uniform motion relative to S will also be inertial for these events. Observers at rest in all such frames will be able to construct rectangular Cartesian axes and synchronize clocks in the manner described in Chapter 1 and hence measure the proper time interval $d\tau$ between the events. If, for one such observer,

GENERAL THEORY OF RELATIVITY 145

the events at the points having rectangular Cartesian coordinates (x,y,z), $(x+dx, y+dy, z+dz)$ occur at the times t, $t+dt$ respectively, then

$$d\tau^2 = dt^2 - \frac{1}{c^2}(dx^2 + dy^2 + dz^2). \tag{46.5}$$

The *interval* between the events ds will be defined by

$$ds^2 = -c^2 d\tau^2 = dx^2 + dy^2 + dz^2 - c^2 dt^2. \tag{46.6}$$

The coordinates (x, y, z, t) of an event in this quasi-inertial frame, will be related to the coordinates x^i defined earlier, by equations

$$x = x(x^1, x^2, x^3, x^4), \text{ etc.} \tag{46.7}$$

and hence

$$dx = \frac{\partial x}{\partial x^i} dx^i, \text{ etc.} \tag{46.8}$$

Substituting for dx, dy, dz, dt in equation (46.6), we obtain the result

$$ds^2 = g_{ij} dx^i dx^j, \tag{46.9}$$

determining the interval ds between two events contiguous in space-time, relative to a general coordinate frame valid for the whole of space-time. The space-time continuum can accordingly be treated as a Riemannian space with metric given by equation (46.9).

47. Motion of a free particle in a gravitational field

In a region of space which is at a great distance from material bodies, rectangular Cartesian axes $Oxyz$ can be found constituting an inertial frame. If time is measured by clocks synchronized within this frame and moving with it, the motion of a freely moving test particle relative to the frame will be uniform. Thus, if (x, y, z) is the position of such a particle at time t, its equations of motion can be written

$$\frac{d^2 x}{dt^2} = \frac{d^2 y}{dt^2} = \frac{d^2 z}{dt^2} = 0. \tag{47.1}$$

Let ds be the interval between the event of the particle arriving at the point (x, y, z) at time t and the contiguous event of the particle arriving at $(x+dx, y+dy, z+dz)$ at $t+dt$. Then ds is given by equation (46.6) and, if v is the speed of the particle, it follows from this equation that

$$ds = (v^2 - c^2)^{1/2} dt. \tag{47.2}$$

Since v is constant, it now follows that equations (47.1) can be expressed in the form

$$\frac{d^2 x}{ds^2} = \frac{d^2 y}{ds^2} = \frac{d^2 z}{ds^2} = 0. \tag{47.3}$$

Also, from equation (47.2), it may be deduced that

$$\frac{d^2 t}{ds^2} = 0. \tag{47.4}$$

Equations (47.3) and (47.4) determine the family of world-lines of free particles in space-time relative to an inertial frame.

Now suppose that any other reference frame and procedure for measuring time is adopted in this region of space, e.g. a frame which is in uniform rotation with respect to an inertial frame might be employed. Let (x^1, x^2, x^3, x^4) be the coordinates of an event in this frame. The interval between two contiguous events will then be given by equation (46.9). If an observer using this frame releases a test particle and observes its motion relative to the frame, he will note that it is not uniform or even rectilinear and will be able to account for this fact by assuming the presence of a gravitational field. He will find that the particle's equations of motion are

$$\frac{d^2 x^i}{ds^2} + \{^i_{jk}\}\frac{dx^j}{ds}\frac{dx^k}{ds} = 0. \tag{47.5}$$

This must be the case for, as shown in section 44, this is a tensor equation defining a geodesic and valid in every frame if it is valid in one. But, in the *xyzt*-frame, the g_{ij} are all constant and the three index symbols vanish. Hence, in this frame, the equations (47.5) reduce to the equations (47.3), (47.4) and these are known to be true for the particle's motion. We have shown, therefore, that the effect of a gravitational field of the reducible variety upon the motion of a test particle can be allowed for when the form taken by the metric

GENERAL THEORY OF RELATIVITY 147

tensor g_{ij} of the space-time manifold is known relative to the frame being employed. This means that the g_{ij} determine, and are determined by, the gravitational field.

The ideas of the previous paragraph will now be extended to regions of space where irreducible gravitational fields are present. It has been pointed out that, for any sufficiently small region of such space and interval of time, an inertial frame can be found and consequently the paths of freely moving particles will be governed in such a small region by equations (47.5). It will now be assumed that these are the equations of motion of free particles without any restriction, i.e. that the world-line of a free particle is a geodesic for the space-time manifold or that the world-line of a free particle has constant direction. This appears to be the natural generalization of the Galilean law of inertia whereby, even in an irreducible gravitational field, a particle's trajectory through space-time is the straightest possible after consideration has been given to the intrinsic curvature of the continuum. It will then follow that the motions of particles falling freely in any gravitational field can be determined relative to any frame when the components g_{ij} of the metric tensor for this frame are known. Thus the g_{ij} will always specify the gravitational field observed to be present in a frame and the only distinction between irreducible and reducible fields will be that, for the latter it will be possible to find a coordinate frame in space-time for which the metric tensor has all its components zero except

$$g_{11} = g_{22} = g_{33} = 1 \quad g_{44} = -c^2, \tag{47.6}$$

whereas for the former this will not be possible.

48. Einstein's law of gravitation

According to Newtonian ideas, the gravitational field which exists in any region of space is determined by the distribution of matter. This suggests that the metric tensor of the space-time manifold, which has been shown to be closely related to the observed gravitational field, should be calculable when the matter distribution throughout space-time is known. We first look, therefore, for a tensor quantity describing this matter distribution with respect to any frame in space-time and then attempt to relate this to the metric tensor. The energy–momentum tensor T_{ij}, defined in Chapter 4 with respect to an inertial frame, immediately suggests itself. Both matter and electromagnetic energy contribute to the components of this tensor but since, according to the special theory, mass and energy are basically identical, it is to be expected that all forms of energy, including the electromagnetic variety, will contribute to the gravitational field.

Since the energy–momentum tensor has been defined in inertial frames only, this definition must now be extended to apply to a general coordinate frame in space-time. This can be carried out thus: A rectangular Cartesian inertial frame can be established for the neighbourhood of any point of a gravitational field and valid for a short time duration. Corresponding to this frame and its associated clocks, rectangular Cartesian axes $Oy^1 y^2 y^3 y^4$ can be constructed in the region surrounding the corresponding point P of space-time. The transformation equations relating the coordinates y^i of an event to its coordinates x^i with respect to any other coordinate frame can now be found. Then, if $T_{ij}^{(0)}$ are the components of the energy–momentum tensor in the y-frame at the point P, its components in the x-frame at this point can be determined from the appropriate tensor transformation equations. Thus, the covariant energy–momentum tensor will have components T_{ij} in the x-frame given by

$$T_{ij} = \frac{\partial y^r}{\partial x^i} \frac{\partial y^s}{\partial x^j} T_{rs}^{(0)}. \tag{48.1}$$

Since covariant and contravariant tensors are indistinguishable with respect to rectangular Cartesian axes, $T_{rs}^{(0)}$ can also be taken to be the components of a contravariant tensor in the y-frame and the components of this tensor in the x-frame will then be given by the equation

$$T^{ij} = \frac{\partial x^i}{\partial y^r} \frac{\partial x^j}{\partial y^s} T_{rs}^{(0)}. \tag{48.2}$$

Similarly, the components of the mixed energy–momentum tensor are given by

$$T_j^i = \frac{\partial x^i}{\partial y^r} \frac{\partial y^s}{\partial x^j} T_{rs}^{(0)}. \tag{48.3}$$

These transformations can be carried out at every point of space-time, thus generating for the x-frame an energy–momentum tensor field throughout the continuum.

Consider the tensor equation

$$T^{ij}{}_{;j} = 0. \tag{48.4}$$

Expressed in terms of the coordinates y^i at any point of space-time, this simplifies to

$$T^{(0)}_{ij,j} = 0, \tag{48.5}$$

which is equation (28.18). Being valid in one frame, therefore, equation (48.4) is true for all frames. Thus, the divergence of the energy–momentum tensor vanishes. If, therefore, this tensor is to be related to the metric tensor g_{ij}, the relationship should be of such a form that it implies equation (48.4). Now

$$g^{ij}{}_{;k} = 0, \tag{48.6}$$

by equation (41.12) and hence, *a fortiori*,

$$g^{ij}{}_{;j} = 0. \tag{48.7}$$

The law
$$T^{ij} = \lambda g^{ij}, \tag{48.8}$$

where λ is a universal constant, would accordingly be satisfactory in this respect. However, over a region in which matter and energy were absent so that $T^{ij} = 0$, this would imply that

$$g^{ij} = 0, \tag{48.9}$$

which is clearly incorrect. Further, according to Newtonian theory, if μ is the density of matter, the gravitational field can be derived from a potential function V satisfying the equation

$$\nabla^2 V = 4\pi\gamma\mu, \tag{48.10}$$

where γ is the gravitational constant. The new law of gravitation which is being sought must include equation (48.10) as an approximation. But, as appears from equation (28.12), T_{44} involves μ and it seems reasonable, therefore, to expect that the other member of the equation expressing the new law of gravitation will provide terms which can receive an approximate interpretation as $\nabla^2 V$. This implies that second order derivatives of the metric tensor components will probably be present. We therefore have a requirement for a second rank contravariant symmetric tensor involving second order derivatives of the g_{ij} and of vanishing divergence to which T^{ij} can be assumed proportional. Einstein's tensor (43.18) possesses these characteristics and consequently we shall put

$$R^{ij} - \tfrac{1}{2} g^{ij} R = -\kappa T^{ij}, \qquad (48.11)$$

where κ is a constant of proportionality which must be related to γ and which we shall later prove to be positive. Equation (48.11) expresses *Einstein's Law of Gravitation*; by lowering the indices successively, it may be expressed in the two alternative forms

$$R^i_j - \tfrac{1}{2} \delta^i_j R = -\kappa T^i_j, \qquad (48.12)$$

$$R_{ij} - \tfrac{1}{2} g_{ij} R = -\kappa T_{ij}. \qquad (48.13)$$

If equation (48.12) is contracted, it is found that

$$R = \kappa T, \qquad (48.14)$$

where $T = T^i_i$. It now follows that Einstein's law of gravitation can also be expressed in the form

$$R_{ij} = \kappa(\tfrac{1}{2} g_{ij} T - T_{ij}), \qquad (48.15)$$

with two other forms obtained by raising subscripts.

Since the divergence of g^{ij} vanishes, a possible alternative to the law (48.11) is

$$R^{ij} - \tfrac{1}{2} g^{ij} R + \lambda g^{ij} = -\kappa T^{ij}, \qquad (48.16)$$

where λ is a constant. The law (48.11) gives results which agree with observation over regions of space of galactic dimensions, so that it is certain that, even if λ is not zero, it is exceedingly small. However, the extra term has entered into some cosmological investigations. It will be disregarded in later sections of this book.

49. Acceleration of a particle in a weak gravitational field

In a gravitational field, such as the one due to the Earth, the geometry of space is not Euclidean and no truly inertial frame exists. In spite of this, we experience no practical difficulty in establishing rectangular

GENERAL THEORY OF RELATIVITY 151

Cartesian axes $Oxyz$ at the Earth's surface relative to which for all practical purposes the geometry is Euclidean and the behaviour of electromagnetic systems is indistinguishable from their behaviour in an inertial frame. It must be concluded, therefore, that such a gravitational field is comparatively weak and hence that, with respect to such axes and their associated clocks, the space-time metric will not differ greatly from that given by equation (46.6). Putting

$$x^1 = x, \quad x^2 = y, \quad x^3 = z, \quad x^4 = ict, \tag{49.1}$$

in terms of the x^i the metric will be given by

$$ds^2 = dx^i dx^i \tag{49.2}$$

approximately. With respect to the x^i-frame, it will accordingly be assumed that

$$g_{ij} = \delta_{ij} + h_{ij}, \tag{49.3}$$

where the δ_{ij} are Kronecker deltas and the h_{ij} are small by comparison.

Consider a particle moving in a weak gravitational field whose metric tensor is given by equation (49.3). The contravariant metric tensor will be given by an equation of the form

$$g^{ij} = \delta^{ij} + k^{ij}, \tag{49.4}$$

where the k^{ij} are of the same order of smallness as the h_{ij}. Then, since

$$[ij,k] = \frac{1}{2}\left(\frac{\partial h_{jk}}{\partial x^i} + \frac{\partial h_{ik}}{\partial x^j} - \frac{\partial h_{ij}}{\partial x^k}\right), \tag{49.5}$$

it follows that, to a first approximation,

$$\{{}^{\ k}_{i\ j}\} = \delta^{kr}[ij,r] = \frac{1}{2}\left(\frac{\partial h_{jk}}{\partial x^i} + \frac{\partial h_{ik}}{\partial x^j} - \frac{\partial h_{ij}}{\partial x^k}\right). \tag{49.6}$$

The equations of motion of the particle can now be written down as at (47.5).

By equation (47.2)

$$\frac{dx^i}{ds} = \frac{dx^i}{dt}\frac{dt}{ds} = (v^2 - c^2)^{-1/2}(\mathbf{v}, ic), \tag{49.7}$$

where $\mathbf{v}$ is the particle's velocity in the quasi-inertial frame. Hence, if the particle is stationary in the frame at the instant under consideration,

$$\frac{dx^i}{ds} = (\mathbf{0}, 1), \quad (49.8)$$

and the equations of motion (47.5) reduce to the form

$$\frac{d^2 x^i}{ds^2} + \{_4{}^i{}_4\} = 0, \quad (49.9)$$

correct to the first order in the h_{ij}. Substituting from equation (49.6), this is seen to be equivalent to

$$\frac{d^2 x^i}{ds^2} = \frac{1}{2}\frac{\partial h_{44}}{\partial x^i} - \frac{\partial h_{4i}}{\partial x^4}. \quad (49.10)$$

Differentiating equation (49.7) with respect to s and making use of equation (47.2), we obtain

$$\frac{d^2 x^i}{ds^2} = (v^2 - c^2)^{-1}\left(\frac{d\mathbf{v}}{dt}, 0\right) - v\frac{dv}{dt}(v^2 - c^2)^{-2}(\mathbf{v}, ic) \quad (49.11)$$

and, when $\mathbf{v} = \mathbf{0}$, this reduces to

$$\frac{d^2 x^i}{ds^2} = -\frac{1}{c^2}\left(\frac{d\mathbf{v}}{dt}, 0\right). \quad (49.12)$$

From equations (49.10), (49.12), we deduce that the components of the acceleration of the stationary particle in the directions of the rectangular axes are

$$-c^2\left(\frac{1}{2}\frac{\partial h_{44}}{\partial x^i} - \frac{\partial h_{4i}}{\partial x^4}\right) \quad (49.13)$$

for $i = 1, 2, 3$. Reverting to the original coordinates (x, y, z, t), these components are written

$$-c^2\left(\frac{1}{2}\frac{\partial h_{44}}{\partial x} + \frac{i}{c}\frac{\partial h_{41}}{\partial t}\right), \text{ etc.} \quad (49.14)$$

Hence, if the field does not vary with the time, the acceleration vector is

$$-\text{grad}(\tfrac{1}{2}c^2 h_{44}). \quad (49.15)$$

GENERAL THEORY OF RELATIVITY 153

But, if V is the Newtonian potential function for the field, this acceleration will be $-\operatorname{grad} V$. It follows that, for a weak field, a Newtonian scalar potential V exists and is related to the space-time metric by the equation

$$V = \tfrac{1}{2}c^2 h_{44}. \tag{49.16}$$

Alternatively, we can write

$$g_{44} = 1 + \frac{2V}{c^2}. \tag{49.17}$$

50. Newton's law of gravitation

In this section it will be shown that Newton's Law of Gravitation may be deduced from Einstein's Law in the normal case when the gravitational field's intensity is weak and the matter distribution is static.

First consider the form taken by the Riemann-Christoffel tensor in the space-time of a weak field. In the x^i-frame, the metric tensor is given by equation (49.3) and the Christoffel three-index symbols by equation (49.6). If products of the h_{ij} are to be neglected, equation (37.21) shows that

$$B^i_{jkl} = \frac{\partial}{\partial x^k}\{^i_{jl}\} - \frac{\partial}{\partial x^l}\{^i_{jk}\} \tag{50.1}$$

approximately. Hence the Ricci tensor is given by

$$\begin{aligned}
R_{jk} &= \frac{\partial}{\partial x^k}\{^i_{ji}\} - \frac{\partial}{\partial x^i}\{^i_{jk}\}, \\
&= \frac{1}{2}\frac{\partial}{\partial x^k}\left\{\frac{\partial h_{ji}}{\partial x^i} + \frac{\partial h_{ii}}{\partial x^j} - \frac{\partial h_{ij}}{\partial x^i}\right\} - \\
&\quad - \frac{1}{2}\frac{\partial}{\partial x^i}\left\{\frac{\partial h_{ij}}{\partial x^k} + \frac{\partial h_{ki}}{\partial x^j} - \frac{\partial h_{jk}}{\partial x^i}\right\}, \\
&= \frac{1}{2}\left\{\frac{\partial^2 h_{ii}}{\partial x^j \partial x^k} + \frac{\partial^2 h_{jk}}{\partial x^i \partial x^i} - \frac{\partial^2 h_{ij}}{\partial x^i \partial x^k} - \frac{\partial^2 h_{ki}}{\partial x^i \partial x^j}\right\}. \tag{50.2}
\end{aligned}$$

In particular, putting $j = k = 4$, we find that

$$R_{44} = \frac{1}{2}\left\{\frac{\partial^2 h_{ii}}{\partial x^4 \partial x^4} + \frac{\partial^2 h_{44}}{\partial x^i \partial x^i} - 2\frac{\partial^2 h_{i4}}{\partial x^i \partial x^4}\right\}. \tag{50.3}$$

L*

If the matter distribution is static in the quasi-inertial frame being employed, the h_{ij} will be independent of t and equation (50.3) reduces to

$$R_{44} = \tfrac{1}{2}\nabla^2 h_{44}, \qquad (50.4)$$

where $\nabla^2 = \partial^2/\partial x^2 + \partial^2/\partial y^2 + \partial^2/\partial z^2$. If V is the Newtonian potential for the field, equation (49.16) now shows that

$$R_{44} = \frac{1}{c^2}\nabla^2 V. \qquad (50.5)$$

Assuming that no electromagnetic field is present, the overall energy–momentum tensor for the mass distribution will be given in the quasi-inertial frame by the equation

$$T_{ij} = \Theta_{ij}, \qquad (50.6)$$

where Θ_{ij} is determined approximately by equation (28.10). Since the distribution is static, its 4-velocity at every point is $(\mathbf{0}, ic)$ and hence all components of T_{ij}, with the exception of T_{44}, are zero. In this case,

$$T_{44} = -c^2 \mu_{00}, \qquad (50.7)$$

where μ_{00}, for zero velocity of matter, is the ordinary mass density. Also

$$T = T^i_i = T_{ii} = T_{44} = -c^2 \mu_{00}. \qquad (50.8)$$

The 44-component of Einstein's gravitation law in the form of equation (48.15) can now be expressed approximately

$$\frac{1}{c^2}\nabla^2 V = \tfrac{1}{2}\kappa c^2 \mu_{00},$$

or
$$\nabla^2 V = \tfrac{1}{2}\kappa c^4 \mu_{00}. \qquad (50.9)$$

This is the Poisson equation (48.10) of classical Newtonian theory, provided we accept

$$\kappa = \frac{8\pi\gamma}{c^4}. \qquad (50.10)$$

This specifies κ in terms of the gravitational constant.

51. Metrics with spherical symmetry

When a change is made in the space-time coordinate frame from coordinates x^i to coordinates $\bar{x}$, the metric tensor g_{ij} will change to $\bar{g}_{ij}$ by the law of transformation of a covariant tensor. In general, the g_{ij} will be functions of the x^i and the $\bar{g}_{ij}$ will be functions of the $\bar{x}^i$, but it will not usually be the case that the $\bar{g}_{ij}$ are the *same* functions of the 'barred' coordinates that the g_{ij} are of the 'unbarred' coordinates. I.e., the functions $g_{ij}(x^k)$ are not *form invariant* under general coordinate transformations. However, in some special cases, it is possible for these functions to be form invariant under a whole group of transformations and we shall study such a case in this section.

In a gravitational field, the geometry can only be quasi-Euclidean and consequently rectangular Cartesian axes do not exist. Nevertheless, no difficulty is experienced in practice in defining such axes approximately and we shall suppose, therefore, that the coordinates x, y, z, t of an event in the gravitational field about to be considered are interpreted physically as rectangular Cartesian coordinates and time. We shall now search for a metric which, when expressed in those coordinates, is form invariant with respect to the group of coordinate transformations which will be interpreted physically as rotations of the rectangular axes $Oxyz$ (t is to remain unaltered). Such a metric will be said to be *spherically symmetric* about O.

Invariants for this group of coordinate transformations, which are of degree no higher than the second in the coordinate differentials dx, dy, dz, are

$$x^2+y^2+z^2, \quad x\,dx+y\,dy+z\,dz, \quad dx^2+dy^2+dz^2. \qquad (51.1)$$

Introducing spherical polar coordinates (r, θ, ϕ), which will be *defined* by the equations (29.3), these invariants may be written

$$r^2, \quad r\,dr, \quad dr^2+r^2\,d\theta^2+r^2\sin^2\theta\,d\phi^2. \qquad (51.2)$$

It follows that
$$r, \quad dr, \quad d\theta^2+\sin^2\theta\,d\phi^2 \qquad (51.3)$$

are invariants. The most general metric with spherical symmetry can now be built up in the form

$$\begin{aligned}ds^2 = {} & A(r,t)\,dr^2 + B(r,t)(d\theta^2+\sin^2\theta\,d\phi^2) + \\ & + C(r,t)\,dr\,dt + D(r,t)\,dt^2.\end{aligned} \qquad (51.4)$$

We now replace r by a new coordinate r' according to the transformation equation

$$r'^2 = B(r,t). \tag{51.5}$$

Then
$$ds^2 = E(r',t)\,dr'^2 + r'^2(d\theta^2 + \sin^2\theta\,d\phi^2) +$$
$$+ F(r',t)\,dr'\,dt + G(r',t)\,dt^2. \tag{51.6}$$

In a truly inertial frame, spherical polar coordinates can be defined exactly and the metric will, by equation (46.6), be expressed in the form

$$ds^2 = dr^2 + r^2(d\theta^2 + \sin^2\theta\,d\phi^2) - c^2\,dt^2. \tag{51.7}$$

Comparing equations (51.6), (51.7), it is clear that, in a region for which (51.6) is the metric, r' will behave physically like a true spherical polar coordinate r. We shall accordingly drop the primes and write

$$ds^2 = E(r,t)\,dr^2 + r^2(d\theta^2 + \sin^2\theta\,d\phi^2) +$$
$$+ F(r,t)\,dr\,dt + G(r,t)\,dt^2. \tag{51.8}$$

If our frame is quasi-inertial, equation (51.7) must be an approximation for equation (51.8) and the following equations must therefore be true approximately:

$$E(r,t) = 1, \quad F(r,t) = 0, \quad G(r,t) = -c^2. \tag{51.9}$$

Consider now the special case when the gravitational field is static in the quasi-inertial frame for which (r,θ,ϕ) are approximate spherical polar coordinates and t is the time. The functions E, F, G will then be independent of t. Also, space-time will be symmetric as regards past and future senses of the time variable and this implies that ds^2 is unaltered when dt is replaced by $-dt$. Thus $F = 0$ and we have

$$ds^2 = a\,dr^2 + r^2(d\theta^2 + \sin^2\theta\,d\phi^2) - bc^2\,dt^2, \tag{51.10}$$

where a, b are functions of r both approximating unity.

For this metric, taking

$$x^1 = r, \quad x^2 = \theta, \quad x^3 = \phi, \quad x^4 = t, \tag{51.11}$$

we have

$$g_{11} = a, \quad g_{22} = r^2, \quad g_{33} = r^2\sin^2\theta, \quad g_{44} = -bc^2, \tag{51.12}$$

all other g_{ij} being zero. Thus

$$g = -abc^2 r^4 \sin^2\theta, \tag{51.13}$$

GENERAL THEORY OF RELATIVITY

and hence

$$g^{11} = \frac{1}{a}, \quad g^{22} = \frac{1}{r^2}, \quad g^{33} = \frac{1}{r^2 \sin^2 \theta}, \quad g^{44} = -\frac{1}{bc^2}, \quad (51.14)$$

all other g^{ij} being zero. The three-index symbols can now be calculated from equation (41.9). Those which do not vanish are listed below:

$$\left.\begin{aligned}
\{_1{}^1{}_1\} &= \frac{a'}{2a}, \\
\{_1{}^2{}_2\} &= \{_2{}^2{}_1\} = \frac{1}{r}, \\
\{_1{}^3{}_3\} &= \{_3{}^3{}_1\} = \frac{1}{r}, \\
\{_1{}^4{}_4\} &= \{_4{}^4{}_1\} = \frac{b'}{2b}, \\
\{_2{}^1{}_2\} &= -\frac{r}{a}, \\
\{_2{}^3{}_3\} &= \{_3{}^3{}_2\} = \cot\theta, \\
\{_3{}^1{}_3\} &= -\frac{r}{a}\sin^2\theta, \\
\{_3{}^2{}_3\} &= -\sin\theta\cos\theta, \\
\{_4{}^1{}_4\} &= \frac{c^2 b'}{2a},
\end{aligned}\right\} \quad (51.15)$$

primes denoting differentiations with respect to r.

Contracting B^i_{jkl} as given by equation (37.21) with respect to the indices i, l, it follows that

$$\begin{aligned}
R_{jk} &= \{_r{}^i{}_k\}\{_i{}^r{}_j\} - \{_r{}^i{}_i\}\{_j{}^r{}_k\} + \frac{\partial}{\partial x^k}\{_j{}^i{}_i\} - \frac{\partial}{\partial x^i}\{_j{}^i{}_k\}, \\
&= \{_r{}^i{}_k\}\{_i{}^r{}_j\} - \{_j{}^r{}_k\}\frac{\partial}{\partial x^r}\log\sqrt{(-g)} + \\
&\quad + \frac{\partial^2}{\partial x^j \partial x^k}\log\sqrt{(-g)} - \frac{\partial}{\partial x^i}\{_j{}^i{}_k\}, \quad (51.16)
\end{aligned}$$

G

by equation (43.4) (g can clearly be replaced by $-g$ in this equation without affecting its validity). The non-zero components of the Ricci tensor are now calculated to be as follows:

$$\left.\begin{aligned}
R_{11} &= \frac{b''}{2b} - \frac{b'^2}{4b^2} - \frac{a'b'}{4ab} - \frac{a'}{ar}, \\
R_{22} &= \frac{rb'}{2ab} - \frac{ra'}{2a^2} + \frac{1}{a} - 1, \\
R_{33} &= R_{22} \sin^2 \theta, \\
R_{44} &= c^2 \left(-\frac{b''}{2a} + \frac{b'^2}{4ab} + \frac{a'b'}{4a^2} - \frac{b'}{ar} \right).
\end{aligned}\right\} \quad (51.17)$$

Since $\quad R_1^1 = g^{11} R_{11}, \quad R_2^2 = g^{22} R_{22}, \text{ etc.} \quad (51.18)$

it will be found that

$$R = R_i^i = \frac{b''}{ab} - \frac{b'^2}{2ab^2} - \frac{a'b'}{2a^2 b} + \frac{2b'}{abr} - \frac{2a'}{a^2 r} + \frac{2}{ar^2} - \frac{2}{r^2}. \quad (51.19)$$

52. Schwarzschild's solution

The static, spherically-symmetrical, metric (51.10) will determine the gravitational field of a static distribution of matter also having spherical symmetry, provided it satisfies Einstein's equations (48.15). We shall consider the special case when the whole of space is devoid of matter, apart from a spherical body with its centre at the centre of symmetry O. Then $T_{ij} = 0$, $T = 0$ at all points outside the body and Einstein's equations reduce in this region to

$$R_{ij} = 0. \quad (52.1)$$

By equations (51.17), these are satisfied by the metric (51.10), provided a, b are such that

$$\frac{b''}{2b} - \frac{b'^2}{4b^2} - \frac{a'b'}{4ab} - \frac{a'}{ar} = 0, \quad (52.2)$$

$$\frac{rb'}{2ab} - \frac{ra'}{2a^2} + \frac{1}{a} - 1 = 0, \quad (52.3)$$

$$-\frac{b''}{2a} + \frac{b'^2}{4ab} + \frac{a'b'}{4a^2} - \frac{b'}{ar} = 0. \quad (52.4)$$

From (52.2), (52.4), it follows that

$$ab' + a'b = 0 \tag{52.5}$$

and hence that $\quad ab = \text{constant.} \tag{52.6}$

But, as $r \to \infty$, we shall assume that our metric approaches that given by equation (51.7) and valid in the absence of a gravitational field. Thus, as $r \to \infty$, then $a \to 1$, $b \to 1$ and hence

$$ab = 1. \tag{52.7}$$

Eliminating b from equation (52.3), it will be found that

$$ra' = a(1-a). \tag{52.8}$$

This equation is easily integrated to yield

$$a = \frac{1}{1-(2m/r)}, \tag{52.9}$$

where m is a constant of integration. Then

$$b = 1 - \frac{2m}{r}, \tag{52.10}$$

and it may be verified that each of the equations (52.2)–(52.4) is satisfied by these expressions for a and b.

We have accordingly arrived at a metric

$$ds^2 = \frac{dr^2}{1-(2m/r)} + r^2(d\theta^2 + \sin^2\theta\, d\phi^2) - c^2\left(1 - \frac{2m}{r}\right)dt^2 \tag{52.11}$$

which is spherically symmetrical and can represent the gravitational field outside a spherical body with its centre at the pole of spherical polar coordinates (r, θ, ϕ). This was first obtained by Schwarzschild. It will be proved in the next section that the constant m is proportional to the mass of the body. This may also be deduced from equation (49.17), for the potential V at a distance r from a spherical body of mass M is given by

$$V = -\frac{\gamma M}{r} \tag{52.12}$$

and hence
$$g_{44} = 1 - \frac{2\gamma M}{c^2 r}. \tag{52.13}$$

Now g_{44} is the coefficient of $(dx^4)^2 = -c^2 dt^2$ in the metric and hence

$$b = 1 - \frac{2\gamma M}{c^2 r}. \tag{52.14}$$

Comparing equations (52.10), (52.14), it will be seen that

$$m = \frac{\gamma M}{c^2}. \tag{52.15}$$

It is clear from equation (52.11) that the metric is singular at a distance

$$r = 2m = \frac{2\gamma M}{c^2} \tag{52.16}$$

from O. We conclude that the radius of the body must certainly be greater than this minimum value. Since, in c.g.s. units, $c = 3 \times 10^{10}$ and for the Earth $\gamma M = 3.991 \times 10^{20}$, the minimum radius for this body is about 9 mm.

53. Planetary orbits

The attractions of the planets upon the Sun cause this body to have a small acceleration relative to an inertial frame. If, therefore, a coordinate frame moving with the Sun is constructed, relative to this frame there will be a gravitational field corresponding to this acceleration in addition to that of the Sun and planets. However, for the purpose of the following analysis, this field and the fields of the planets will be neglected. Thus, relative to spherical polar coordinates having their pole at the centre of the Sun, the gravitational field will be assumed determined by the Schwarzschild metric (52.11). The planets will be treated as particles possessing negligible gravitational fields, whose world-lines are geodesics in space-time (section 47). We proceed to calculate these geodesics.

GENERAL THEORY OF RELATIVITY 161

Substituting for a, b from equations (52.9), (52.10) in equations (51.15), it will be found that

$$\left.\begin{aligned}
\{_1^1{}_1\} &= -\frac{m}{r(r-2m)}, \\
\{_1^2{}_2\} &= \frac{1}{r}, \\
\{_1^3{}_3\} &= \frac{1}{r}, \\
\{_1^4{}_4\} &= \frac{m}{r(r-2m)}, \\
\{_2^1{}_2\} &= -(r-2m), \\
\{_2^3{}_3\} &= \cot\theta, \\
\{_3^1{}_3\} &= -(r-2m)\sin^2\theta, \\
\{_3^2{}_3\} &= -\sin\theta\cos\theta, \\
\{_4^1{}_4\} &= \frac{mc^2}{r^3}(r-2m).
\end{aligned}\right\} \quad (53.1)$$

Equations (44.5) for the geodesics accordingly take the form:

$$\frac{d^2r}{ds^2} - \frac{m}{r(r-2m)}\left(\frac{dr}{ds}\right)^2 - (r-2m)\left\{\left(\frac{d\theta}{ds}\right)^2 + \sin^2\theta\left(\frac{d\phi}{ds}\right)^2 - \frac{mc^2}{r^3}\left(\frac{dt}{ds}\right)^2\right\} = 0, \quad (53.2)$$

$$\frac{d^2\theta}{ds^2} + \frac{2}{r}\frac{dr}{ds}\frac{d\theta}{ds} - \sin\theta\cos\theta\left(\frac{d\phi}{ds}\right)^2 = 0, \quad (53.3)$$

$$\frac{d^2\phi}{ds^2} + \frac{2}{r}\frac{dr}{ds}\frac{d\phi}{ds} + 2\cot\theta\frac{d\theta}{ds}\frac{d\phi}{ds} = 0, \quad (53.4)$$

$$\frac{d^2t}{ds^2} + \frac{2m}{r(r-2m)}\frac{dr}{ds}\frac{dt}{ds} = 0. \quad (53.5)$$

There is also available the first integral (44.6) of these equations which, by equation (52.11), is

$$\frac{r}{r-2m}\left(\frac{dr}{ds}\right)^2 + r^2\left\{\left(\frac{d\theta}{ds}\right)^2 + \sin^2\theta\left(\frac{d\phi}{ds}\right)^2\right\} - \frac{c^2}{r}(r-2m)\left(\frac{dt}{ds}\right)^2 = 1. \quad (53.6)$$

Equation (53.2) will be discarded in favour of this first integral.

We now choose the spherical polar coordinates so that the planet is moving initially in the plane $\theta = \tfrac{1}{2}\pi$. Then $d\theta/ds = 0$ initially and hence, by equation (53.3), $d^2\theta/ds^2 = 0$ at this instant and the particle continues to move in this plane indefinitely. Thus, putting $\theta = \tfrac{1}{2}\pi$, $d\theta/ds = 0$ in the remaining geodesic equations, we find these reduce to

$$\frac{d^2\phi}{ds^2} + \frac{2}{r}\frac{dr}{ds}\frac{d\phi}{ds} = 0, \quad (53.7)$$

$$\frac{d^2t}{ds^2} + \frac{2m}{r(r-2m)}\frac{dr}{ds}\frac{dt}{ds} = 0, \quad (53.8)$$

$$\frac{r}{r-2m}\left(\frac{dr}{ds}\right)^2 + r^2\left(\frac{d\phi}{ds}\right)^2 - \frac{c^2}{r}(r-2m)\left(\frac{dt}{ds}\right)^2 = 1. \quad (53.9)$$

Putting $w = d\phi/ds$, $v = dt/ds$, equations (53.7), (53.8) may be written

$$\frac{dw}{dr} + \frac{2}{r}w = 0, \quad (53.10)$$

$$\frac{dv}{dr} + \frac{2m}{r(r-2m)}v = 0, \quad (53.11)$$

respectively. These equations can now be integrated to yield

$$w = \frac{d\phi}{ds} = \frac{\alpha}{r^2}, \quad (53.12)$$

$$v = \frac{dt}{ds} = \frac{\beta r}{r-2m}, \quad (53.13)$$

where α, β are constants of integration.

Substituting for $d\phi/ds$, dt/ds from the last two equations into equation (53.9), it follows that

$$\left(\frac{dr}{ds}\right)^2 + \frac{\alpha^2}{r^3}(r-2m) = 1 + c^2\beta^2 - \frac{2m}{r}. \tag{53.14}$$

Then, eliminating ds between this equation and equation (53.12), we obtain the equation for the orbit, viz.

$$\left(\frac{\alpha}{r^2}\frac{dr}{d\phi}\right)^2 + \frac{\alpha^2}{r^2} = 1 + c^2\beta^2 - \frac{2m}{r} + \frac{2m\alpha^2}{r^3}. \tag{53.15}$$

With $u = 1/r$, this reduces to the form

$$\left(\frac{du}{d\phi}\right)^2 + u^2 = \frac{1+c^2\beta^2}{\alpha^2} - \frac{2m}{\alpha^2}u + 2mu^3. \tag{53.16}$$

Differentiating through with respect to ϕ, this equation takes a form which is familiar in the theory of orbits, viz.

$$\frac{d^2u}{d\phi^2} + u = -\frac{m}{\alpha^2} + 3mu^2. \tag{53.17}$$

The corresponding equation governing the orbit according to classical mechanics is

$$\frac{d^2u}{d\phi^2} + u = \frac{\gamma M}{h^2}, \tag{53.18}$$

where M is the mass of the attracting body and h is the constant velocity moment of the planet about the centre of attraction, i.e.,

$$r^2\frac{d\phi}{dt} = h. \tag{53.19}$$

The general relativity counterpart of this last equation is clearly equation (53.12). If τ is the proper time for the orbiting body, by equation (46.6) $ic\,d\tau = ds$ and equation (53.12) is equivalent to

$$r^2\frac{d\phi}{d\tau} = ic\alpha. \tag{53.20}$$

We shall identify the time variable t of classical mechanics with the proper time τ for the body whose motion is being studied and then, by comparison of equations (53.19), (53.20), it is seen that

$$h = ic\alpha. \tag{53.21}$$

Equation (53.17) then becomes

$$\frac{d^2u}{d\phi^2} + u = \frac{mc^2}{h^2} + 3mu^2 \tag{53.22}$$

which, apart from a term $3mu^2$, is identical with the classical equation (53.18) provided

$$m = \frac{\gamma M}{c^2}. \tag{53.23}$$

This confirms equation (52.15).

The ratio of the additional term $3mu^2$ to the 'inverse square law' term mc^2/h^2 is

$$\frac{3h^2u^2}{c^2} = \frac{3}{c^2}r^2\dot{\phi}^2 \tag{53.24}$$

by equation (53.19). $r\dot{\phi}$ is the transverse component of the planet's velocity and, for the planets of the solar system, takes its largest value in the case of Mercury, viz. 48 km/sec. Since $c = 3 \times 10^5$ km/sec, the ratio of the terms is in this case $7 \cdot 7 \times 10^{-8}$, which is very small. However, the effect of the additional term proves to be cumulative, as will now be proved, and for this reason an observational check can be made.

The solution of the classical equation (53.18), viz.

$$u = \frac{\mu}{h^2}\{1 + e\cos(\phi - \tilde{\omega})\}, \tag{53.25}$$

where $\mu = \gamma M = mc^2$, e is the eccentricity of the orbit and $\tilde{\omega}$ is the longitude of perihelion, will be an approximate, though highly accurate, solution of equation (53.22). Hence the error involved in taking

$$3mu^2 = \frac{3m\mu^2}{h^4}\{1 + e\cos(\phi - \tilde{\omega})\}^2 \tag{53.26}$$

will be absolutely inappreciable, since this term is very small in any case. Equation (53.22) can accordingly be replaced by

$$\frac{d^2u}{d\phi^2}+u = \frac{\mu}{h^2}+\frac{3m\mu^2}{h^4}\{1+e\cos(\phi-\tilde{\omega})\}^2. \qquad (53.27)$$

This equation will possess a solution of the form (53.25) with additional 'particular integral' terms corresponding to the new term (53.26). These prove to be as follows:

$$\frac{3m\mu^2}{h^4}\{1+\tfrac{1}{2}e^2-\tfrac{1}{6}e^2\cos 2(\phi-\tilde{\omega})+e\phi\sin(\phi-\tilde{\omega})\}. \qquad (53.28)$$

The constant term cannot be observationally separated from that already occurring in equation (53.25). The term in $\cos 2(\phi-\tilde{\omega})$ has amplitude too small for detection. However, the remaining term has an amplitude which increases with ϕ and its effect is accordingly cumulative. Adding this to the solution (53.25), we obtain

$$u = \frac{\mu}{h^2}\left\{1+e\cos(\phi-\tilde{\omega})+\frac{3m\mu e}{h^2}\phi\sin(\phi-\tilde{\omega})\right\},$$

$$= \frac{\mu}{h^2}\{1+e\cos(\phi-\tilde{\omega}-\delta\tilde{\omega})\}, \qquad (53.29)$$

where $\delta\tilde{\omega} = 3m\mu\phi/h^2$ and we have neglected terms $O(\delta\tilde{\omega}^2)$.

Equation (53.29) indicates that the longitude of perihelion should steadily increase according to the equation

$$\delta\tilde{\omega} = \frac{3m\mu}{h^2}\phi = \frac{3\mu^2}{c^2h^2}\phi = \frac{3\mu}{c^2l}\phi, \qquad (53.30)$$

where $l = h^2/\mu$ is the semi-latus rectum of the orbit. Taking

$$\mu = 1\cdot 33\times 10^{26}$$

c.g.s. units for the Sun, $c = 3\times 10^{10}$ cm/sec and $l = 5\cdot 79\times 10^{12}$ cm for Mercury, it will be found that the predicted angular advance of perihelion per century for this planet's orbit is 43″. This is in agreement with the observed value. The advances predicted for the other planets are too small to be observable at the present time.

54. Gravitational deflection of a light ray

In section 7 it was shown that the proper time interval between the transmission of a light signal and its reception at a distant point is zero. It was there assumed that the signal was being propagated in an inertial frame and hence that no gravitational field was present. This result can be expressed by saying that

$$ds = 0 \tag{54.1}$$

for any two neighbouring points on the world-line of a light signal. Now, null-geodesics in the space-time having metric (46.6) are defined by equation (54.1) and the equations

$$\frac{d^2 x}{d\lambda^2} = \frac{d^2 y}{d\lambda^2} = \frac{d^2 z}{d\lambda^2} = \frac{d^2 t}{d\lambda^2}, \tag{54.2}$$

for the three index symbols are all zero. Equations (54.2) imply that along a null-geodesic x, y, z are linearly dependent upon t. But this is certainly true for the coordinates of a light signal being propagated in an inertial frame. We conclude that the world-lines of light signals are null-geodesics in space-time.

Since an inertial frame can always be found for a sufficiently small space-time region even in the presence of a gravitational field, it follows that the world-line of a light signal in any such region is a null-geodesic. We shall accept the obvious generalization of this result, viz. that the world-lines of light signals over an unlimited region of space-time are null-geodesics.

We shall now employ this principle to calculate the path of a light ray in the gravitational field of a spherical body. Taking the space-time metric in the Schwarzschild form (equation (52.11)), the three-index symbols are given at equation (53.1) and the equations governing a null-geodesic (equations (44.14)) are identical with the equations (53.2)–(53.5) after s has been replaced by λ. The first integral (44.13) takes the form

$$\frac{r}{r-2m}\left(\frac{dr}{d\lambda}\right)^2 + r^2\left\{\left(\frac{d\theta}{d\lambda}\right)^2 + \sin^2\theta\left(\frac{d\phi}{d\lambda}\right)^2\right\} - \frac{c^2}{r}(r-2m)\left(\frac{dt}{d\lambda}\right)^2 = 0. \tag{54.3}$$

Without loss of generality, we shall again put $\theta = \frac{1}{2}\pi$, so that a ray in the equatorial plane is being considered and then proceed exactly as in the last section to derive the equation

$$\frac{d^2u}{d\phi^2} + u = 3mu^2, \tag{54.4}$$

where $u = 1/r$. This equation determines the family of light rays in the equatorial plane.

As a first approximation to the solution of equation (54.4), we shall neglect the right-hand member. Then

$$u = \frac{1}{R}\cos(\phi + \alpha), \tag{54.5}$$

where R, α are constants of integration. This is the polar equation of a straight line whose perpendicular distance from the centre of attraction is R. As might have been expected, therefore, provided the gravitational field is not too intense, the light rays will be straight lines. This deduction is, of course, confirmed by observation. Thus, as the Moon's motion causes its disc to approach the position of a star on the celestial sphere and ultimately to occult this body, no appreciable deflection of the position of the star on the celestial sphere can be detected.

Again, without loss of generality, we shall put $\alpha = 0$ so that the light ray, as given by equation (54.5), is parallel to the y-axis ($\phi = \pm \frac{1}{2}\pi$). Then, putting $u = \cos\phi/R$ in the right-hand member of equation (54.4), this becomes

$$\frac{d^2u}{d\phi^2} + u = \frac{3m}{R^2}\cos^2\phi. \tag{54.6}$$

The additional 'particular integral' term is now found to be

$$\frac{m}{R^2}(2 - \cos^2\phi) \tag{54.7}$$

and hence the second approximation to the polar equation of the light ray is

$$u = \frac{1}{R}\cos\phi + \frac{m}{R^2}(2 - \cos^2\phi). \tag{54.8}$$

At each end of the ray $u = 0$ and hence

$$\frac{m}{R}\cos^2\phi - \cos\phi - \frac{2m}{R} = 0. \qquad (54.9)$$

Assuming m/R to be small, this quadratic equation has a small root and a large root. The small root is approximately

$$\cos\phi = -\frac{2m}{R} \qquad (54.10)$$

and hence

$$\phi = \pm\left(\frac{\pi}{2} + \frac{2m}{R}\right) \qquad (54.11)$$

at the two ends of the ray. The angular deflection in the ray caused by its passage through the gravitational field is accordingly

$$\frac{4m}{R} \qquad (54.12)$$

approximately.

For a light ray grazing the Sun's surface,

R = Sun's radius = 6.95×10^{10} cm and $m = 1.5 \times 10^5$ cm.

Thus the predicted deflection is 8.62×10^{-6} radians, or about $1.77''$. This prediction has been checked by observing a star close to the Sun's disc during a total eclipse. The experimental findings are in accord with the theoretical result.

55. Gravitational displacement of spectral lines

A *standard clock* will be taken to be any device which experiences a periodic motion, each cycle of which is indistinguishable from every other cycle. The passage of time between two events which occur in the neighbourhood of the clock is then measured by the number of cycles and fraction of a cycle which the device completes between these two instants. The clocks employed to determine the time coordinate ξ^4 of an event in section 46 were not, necessarily, standard clocks. Such coordinate clocks can have arbitrary variable rates, the only requirement being that, if A, B are two events in the vicinity of

GENERAL THEORY OF RELATIVITY   169

a coordinate clock and *B* occurs after *A*, then the coordinate-time for *B* must be greater than the coordinate-time for *A*.

Consider, then, a standard clock which is moving in any manner within some reference frame. Let *A*, *B* be the two events representing the commencement and termination of one clock cycle and let *C*, *D* be two events separated by another clock cycle. Since we are assuming that the clock cycles are indistinguishable, the geometrical relationship between *A* and *B* in space-time must be identical with the relationship between *C* and *D*. It follows that the space-time intervals between *A* and *B* and between *C* and *D* are equal. Thus, every cycle of a standard clock registers its advance along its world-line through a constant interval and, if *ds* is the interval between neighbouring points on this world-line, the quantity whose passage is registered by the clock will be

$$\int ds, \qquad (55.1)$$

integrated along the world-line. I.e. a standard clock registers the passage of interval along its trajectory.

Let x^i ($i = 1, 2, 3, 4$) be the coordinates of an event with respect to some space-time reference frame, x^1, x^2, x^3 being interpreted physically as spatial coordinates relative to a static frame and x^4/ic as time. If a standard clock is at rest relative to this frame, for adjacent points on its world-line $dx^1 = dx^2 = dx^3 = 0$ and hence

$$ds^2 = g_{44}(dx^4)^2 = -c^2 g_{44} dt^2, \qquad (55.2)$$

where we have put $x^4 = ict$. The interval *s* measured by the clock is therefore related to *t*, the coordinate-time at the point (x^1, x^2, x^3), by the equation

$$s = ic \int \sqrt{(g_{44})} \, dt. \qquad (55.3)$$

s is imaginary only because we have chosen to define the interval *ds* between two events in such a way that time-like intervals are imaginary. The standard clock can be graduated to register $T = s/ic$ and then standard clock time *T* will be related to coordinate-time by the equation

$$T = \int \sqrt{(g_{44})} \, dt. \qquad (55.4)$$

In the special case of the coordinate frame employed in section 49 which was stationary in a relatively weak static gravitational field, it was proved that g_{44} is given in terms of the Newtonian scalar potential V for the field by the approximate equation (49.17). Thus

$$dT = \left(1 + \frac{2V}{c^2}\right)^{1/2} dt \tag{55.5}$$

relates time intervals measured by a stationary standard clock and a coordinate-clock at a point in a gravitational field where the potential is V. Now, when it is emitting its characteristic spectrum, an atom is operating as a standard clock. If, therefore, two similar atoms are stationary at different positions in a static gravitational field and they emit their characteristic radiation, the two intervals corresponding to the emission of one complete cycle of a certain spectral line by each atom will be identical. If dT is this standard-time interval, V_1, V_2 are the gravitational potentials at the atoms, and dt_1, dt_2 are the periods of the complete cycles as measured in coordinate-time, then by equation (55.5)

$$dT = \left(1 + \frac{2V_1}{c^2}\right)^{1/2} dt_1 = \left(1 + \frac{2V_2}{c^2}\right)^{1/2} dt_2 \tag{55.6}$$

and hence $\quad dt_1 : dt_2 = \left(1 + \frac{2V_2}{c^2}\right)^{1/2} : \left(1 + \frac{2V_1}{c^2}\right)^{1/2}. \tag{55.7}$

Suppose that the radiation from the atoms is observed at some point P in the field. Let t_a be the coordinate-time at one atom when a radiation wave crest is emitted and t_b the corresponding time for the next crest. Let t'_a, t'_b be the respective coordinate-times of arrival of these crests at P. Since the field and coordinate frame are both static, the time delay between a crest leaving the atom and its arrival at P will be constant. Thus

$$t'_a - t_a = t'_b - t_b, \tag{55.8}$$

or $\quad t'_a - t'_b = t_a - t_b. \tag{55.9}$

This last equation shows that the period of vibration of the atom as measured by the coordinate-clock at P is independent of this point's position and is equal to the period as measured by the coordinate-clock at the atom itself. The frequencies ν_1, ν_2 of corresponding spectral lines for the two atoms as measured at P will therefore, by equation (55.7), be in the ratio

$$\frac{\nu_1}{\nu_2} = \sqrt{\left(\frac{1+2V_1/c^2}{1+2V_2/c^2}\right)} = 1 + \frac{V_1 - V_2}{c^2} \qquad (55.10)$$

approximately.

In the case of an atom at the surface of the Sun and a similar atom at the Earth's surface, it will be found that, in c.g.s. units,

$$V_1 = -9\cdot512 \times 10^{12} \quad \text{(Earth)},$$
$$V_2 = -1\cdot914 \times 10^{15} \quad \text{(Sun)},$$

and thus
$$\frac{\nu_1}{\nu_2} = 1\cdot00000212. \qquad (55.11)$$

This effect is so small, that it is very difficult to measure. However, in the case of the companion of Sirius, the predicted effect is 30 times larger and has been confirmed by observation.

56. Maxwell's equations in a gravitational field

In this final section, the equations (23.11) determining the electromagnetic field due to the motion in vacuo of a distribution of electric charge, will be generalized to take account of any gravitational field which may be present, but we shall not elaborate upon the implications of the modified equations.

Over any sufficiently small region of space and restricted interval of time it is possible to define a rectangular Cartesian inertial frame, i.e. the frame in 'free fall' in the gravitational field. If the electric and magnetic components of the electromagnetic field are measured in this frame, the field tensor F_{ij} defined by equation (23.5) can be found. Employing the appropriate transformation equations, the components of this tensor relative to general coordinates x^i in the gravitational field can be computed. No distinction is made between covariant and contravariant properties relative to the original inertial frame so that, when transforming, F_{ij} may be treated as a covariant, contravariant or mixed tensor. If it is treated as a covariant tensor, the covariant components F_{ij} in the general x^i-frame will be generated. If it is treated as a contravariant or as a mixed tensor, the contravariant or mixed components F^{ij}, F^i_j respectively will be generated. In this way, the field tensor is defined at every point of space-time. Similarly, a current density vector with covariant components J_i and contravariant components J^i is defined relative to the x^i-frame.

Consider the equations

$$F^{ij}{}_{;j} = \frac{4\pi}{c} J^i, \qquad (56.1)$$

$$F_{ij;k} + F_{jk;i} + F_{ki;j} = 0. \qquad (56.2)$$

These are tensor equations and hence are valid in every space-time frame if they are valid in any one. But, relative to the inertial coordinate frame (x, y, z, ict) which can be found for any sufficiently small space-time region, these equations reduce to equations (23.11) and hence are valid over such a region. Regarding the whole of space-time as an aggregate of such small elements, it follows that equations (56.1), (56.2) are universally true.

Since F^{ij} is skew-symmetric,

$$\begin{aligned} F^{ij}{}_{;j} &= \frac{\partial F^{ij}}{\partial x^j} + \{{}^i_{rj}\} F^{rj} + \{{}^j_{rj}\} F^{ir}, \\ &= \frac{\partial F^{ij}}{\partial x^j} + \frac{1}{\sqrt{(-g)}} \frac{\partial}{\partial x^r} \{\sqrt{(-g)}\} F^{ir}, \\ &= \frac{1}{\sqrt{(-g)}} \frac{\partial}{\partial x^j} \{\sqrt{(-g)} F^{ij}\}, \qquad (56.3) \end{aligned}$$

by equation (43.4) (g has been replaced by $-g$, since g is always negative for a real gravitational field). Equation (56.1) is accordingly equivalent to

$$\frac{1}{\sqrt{(-g)}} \frac{\partial}{\partial x^j} \{\sqrt{(-g)} F^{ij}\} = \frac{4\pi}{c} J^i. \qquad (56.4)$$

Further, since g is a relative invariant of weight 2 (cf. equation (32.7)) and hence $\sqrt{(-g)}$ is an invariant density, it follows that $\mathfrak{F}^{ij}, \mathfrak{J}^i$ defined by the equations

$$\mathfrak{F}^{ij} = \sqrt{(-g)}\,F^{ij}, \quad \mathfrak{J}^i = \sqrt{(-g)}\,J^i, \qquad (56.5)$$

are densities and then equation (56.4) takes the simpler form

$$\frac{\partial \mathfrak{F}^{ij}}{\partial x^j} = \frac{4\pi}{c}\mathfrak{J}^i. \qquad (56.6)$$

Also, in view of the skew-symmetry of the field tensor, it follows that equation (56.2) is equivalent to

$$\frac{\partial F_{ij}}{\partial x^k} + \frac{\partial F_{jk}}{\partial x^i} + \frac{\partial F_{ki}}{\partial x^j} = 0. \qquad (56.7)$$

Exercises 6

1. Prove that when an index of T^{ij} as defined by equation (48.2) is lowered, T^i_j as defined by equation (48.3) is obtained.

2. In the space-time whose metric is given by

$$ds^2 = e^{2\phi}(dx^4)^2 + e^{2\theta}(dx^1)^2 + (dx^2)^2 + (dx^3)^2,$$

where ϕ, θ are functions of x^1 only, prove that the Riemann-Christoffel tensor vanishes if and only if

$$\phi'' - \theta'\phi' + \phi'^2 = 0$$

where the dashes denote differentiations with respect to x^1. If $\phi = -\theta$, prove that the space is flat provided that

$$\phi = \tfrac{1}{2}\log(a + bx^1),$$

where a, b are constants.

(L.U.)

3. If the metric of space-time is

$$ds^2 = -e^{\lambda}\{(dx^1)^2 + (dx^2)^2\} - (x^2)^2 e^{-\rho}(dx^3)^2 + e^{\rho}(dx^4)^2,$$

where λ and ρ are functions of x^1 and x^2 only, show, by calculating R_{44}, that the field equations $R_{ij} = 0$ (for a region devoid of matter) require that ρ shall satisfy

$$\frac{\partial^2 \rho}{(\partial x^1)^2} + \frac{\partial^2 \rho}{(\partial x^2)^2} + \frac{1}{x^2}\frac{\partial \rho}{\partial x^2} = 0.$$

(Li.U.)

4. Find the differential equations of the paths of test particles in the space-time of which the metric is

$$ds^2 = e^{2kx}[-(dx^2+dy^2+dz^2)+dt^2],$$

where k is a constant. If

$$v^2 = \left(\frac{dx}{dt}\right)^2 + \left(\frac{dy}{dt}\right)^2 + \left(\frac{dz}{dt}\right)^2$$

and if $v = V$ when $x = 0$, show that

$$1-v^2 = (1-V^2)e^{2kx}.$$

(Li. U.)

5. Use the equations

$$R_j^i - \tfrac{1}{2}\delta_j^i R = -\kappa T_j^i$$

to find the energy–momentum tensor for the distribution of matter corresponding to the space-time

$$ds^2 = -e^g(dx^2+dy^2+dz^2)+dt^2,$$

where g is a function of t only.

(Li. U.)

6. If

$$ds^2 = \frac{dt^2}{1-kx} - \frac{1}{c^2}\frac{dx^2+dy^2+dz^2}{(1-kx)^2}$$

where k is a constant, and if

$$v^2 = \left(\frac{dx}{dt}\right)^2 + \left(\frac{dy}{dt}\right)^2 + \left(\frac{dz}{dt}\right)^2,$$

prove that, along a geodesic,

$$V^2 - v^2 = kc^2 x,$$

where V is a constant.

(L. U.)

7. Show that the four differential equations (53.2)–(53.5) for the geodesics in the Schwarzschild space-time have a solution for which

$\theta = \frac{1}{2}\pi, r = a$, where a is a constant greater than $3m$, and that the total interval along this geodesic from $\phi = 0$ to $\phi = 2\pi$ is

$$2\pi i a \left(\frac{a}{m} - 3\right)^{1/2}.$$

Also show that there is a geodesic along which $\theta = $ const., $\phi = $ const. and which satisfies an equation of the form

$$\left(\frac{dr}{dt}\right)^2 = 2mc^2 \left(1 - \frac{2m}{R}\right)^{-1} \left(1 - \frac{2m}{r}\right)^2 \left(\frac{1}{r} - \frac{1}{R}\right),$$

where R is fixed. State briefly the physical interpretation of these results.

(L. U.)

8. A space-time has metric

$$ds^2 = e^{2\sigma}\{(dx^1)^2 + (dx^2)^2 + (dx^3)^2 + (dx^4)^2\}$$

where σ is a function of (x^1, x^2, x^3, x^4). If t^i is the unit tangent to a geodesic, prove that

$$\frac{dt^k}{ds} + 2(\sigma_i t^i) t^k = \sigma_k e^{-2\sigma},$$

where $$\sigma_k = \frac{\partial \sigma}{\partial x^k}.$$

For slow motions in slowly changing fields, prove that the geodesics are paths of particles in a gravitational field of potential $-\sigma c^2$.

(L. U).

9. Find the Riemann-Christoffel tensor of the space-time of the last exercise, and prove that the scalar curvature R vanishes if, and only if,

$$\sigma_{pp} + \sigma_p \sigma_p = 0$$

where $$\sigma_{pq} = \frac{\partial^2 \sigma}{\partial x^p \partial x^q}.$$

If σ is a function of $r = [(x^1)^2 + (x^2)^2 + (x^3)^2]^{1/2}$ only, prove that this condition is

$$\sigma'' + \frac{2}{r}\sigma' + \sigma'^2 = 0,$$

where dashes denote differentiations with respect to r.

(L.U.)

10. In the space-time of metric

$$ds^2 = e^{2\sigma}\{dr^2 + r^2 d\theta^2 + r^2 \sin^2\theta \, d\phi^2 - dt^2\}$$

where $\sigma = \log(1 + m/r)$, and m is constant, prove that the scalar curvature is zero (see Exercise 9 for this condition). Prove that the geodesics satisfy the equations

$$r^2 \sin^2\theta \frac{d\phi}{ds} = k_1 e^{-2\sigma}$$

$$\frac{dt}{ds} = k_2 e^{-2\sigma},$$

where k_1, k_2 are constants. If we choose $\phi = 0$, $d\phi/ds = 0$ initially, prove that ϕ is always zero, and that

$$r^2 \frac{d\theta}{ds} = h e^{-2\sigma},$$

where h is constant.

(L.U.)

11. Show that

$$ds^2 = e^{-2q(\sigma)}\left\{dt^2 - \frac{1}{c^2}(dx^2 + dy^2 + dz^2)\right\}$$

where q is an arbitrary function and

$$\sigma^2 = t^2 - \frac{1}{c^2}(x^2 + y^2 + z^2)$$

is invariant under a Lorentz transformation.

If

$$j^1 = \rho x, \quad j^2 = \rho y, \quad j^3 = \rho z, \quad j^4 = \rho t$$

where ρ is a function of σ and j^r is a contravariant vector satisfying

$$\frac{\partial}{\partial x^r}(\sqrt{(-g)}\,j^r) = 0, \quad (x^1 = x,\ldots,x^4 = t),$$

show that

$$\rho = \frac{A\,e^{4q}}{\sigma^4},$$

where A is a constant.

(L.U.)

12. By replacing the spherical polar coordinate r occurring in the Schwarzschild metric (52.11) by a new coordinate r' where

$$r = r'\left(1+\frac{m}{2r'}\right)^2,$$

obtain this metric in 'isotropic' form, viz.

$$ds^2 = \left(1+\frac{m}{2r'}\right)^4 (dr'^2 + r'^2\,d\theta^2 + r'^2 \sin^2\theta\,d\phi^2) - \left(\frac{1-m/2r'}{1+m/2r'}\right)^2 c^2\,dt^2.$$

13. Employing a certain frame, an event is specified by spatial coordinates (x,y,z) and a time t. The corresponding space-time manifold has metric

$$ds^2 = dx^2 + dy^2 + dz^2 + 2at\,dx\,dt - (c^2 - a^2 t^2)\,dt^2.$$

Show that a particle falling freely in the gravitational field observed in the frame has equations of motion

$$x = A + Bt - \tfrac{1}{2}at^2, \quad y = C + Dt, \quad z = E + Ft,$$

where A, B, C, D, E, F are constants. By transforming to coordinates (x', y, z, t), where $x' = x + \tfrac{1}{2}at^2$, and recalculating the metric, explain this result.

14. (x^1, x^2, x^3) are spatial coordinates of an event relative to a frame S and x^4 is the time of the event measured by a clock in S. A second frame I is falling freely in the neighbourhood of P and may be regarded as inertial. $Oy^1 y^2 y^3$ are rectangular cartesian axes in I and y^4/ic represents the time within I as measured by synchronised

clocks attached to the frame. Show that g_{ij}, the metric tensor in S, is given by

$$g_{ij} = \frac{\partial y^k}{\partial x^i}\frac{\partial y^k}{\partial x^j}.$$

P is a point, fixed in S, having coordinates (x^1, x^2, x^3). At the instant x^4, I is chosen so that P is instantaneously at rest in I. Deduce that

$$\frac{\partial y^4}{\partial x^i} = \frac{g_{i4}}{\sqrt{(g_{44})}}.$$

dl is the distance between P and a neighbouring point

$$P'(x^1+dx^1, x^2+dx^2, x^3+dx^3)$$

as measured by a standard rod in I at the instant x^4. Prove that

$$dl^2 = dy^\alpha dy^\alpha = \gamma_{\lambda\mu} dx^\lambda dx^\mu,$$

where α, λ, μ range over the values 1, 2, 3 and

$$\gamma_{\lambda\mu} = g_{\lambda\mu} - \frac{g_{\lambda 4} g_{\mu 4}}{g_{44}}.$$

($\gamma_{\lambda\mu}$ is the metric tensor for the $\mathscr{R}_3$ which is S at the instant x_4.)

15. $Oxyz$ is a rectangular cartesian inertial frame I. A rigid disc rotates in the xy-plane about its centre O with angular velocity ω. Polar coordinates (r, θ) in a frame R rotating with the disc are defined by the equations

$$x = r\cos(\theta+\omega t), \quad y = r\sin(\theta+\omega t),$$

where t is the time measured by synchronised clocks in the inertial frame. If the time of an event in R is taken to be the time shown by an adjacent clock in I, show that the space-time metric associated with R is

$$ds^2 = dr^2 + r^2 d\theta^2 + 2\omega r^2 d\theta\, dt - (c^2 - r^2\omega^2)\, dt^2.$$

Deduce that the metric for geometry in R is given by

$$dl^2 = dr^2 + \frac{r^2 d\theta^2}{1 - \omega^2 r^2/c^2}.$$

(Hint: employ the result of the previous exercise.) Hence show that the family of geodesics on the disc is determined by the equation

$$\theta = \text{const.} - \sin^{-1}\left(\frac{a}{r}\right) - \frac{a}{r^2}\sqrt{(r^2 - a^2)},$$

where $r_1 = c/\omega$ and $|a| < r_1$. Sketch this family. What is the physical significance of r_1?

16. x^i ($i = 1, 2, 3, 4$) are three space coordinates and time relative to a reference frame S. A test particle is momentarily at rest in S at the point (x^1, x^2, x^3) at the time x^4. If g_{ij} is the metric tensor for the gravitational field in S, write down the conditions that the world-line of the particle is a geodesic and deduce that

$$g_{i\alpha}\frac{d^2 x^\alpha}{(dx^4)^2} = \frac{1}{2}\left(\frac{\partial g_{44}}{\partial x^i} + \frac{g_{i4}}{g_{44}}\frac{\partial g_{44}}{\partial x^4}\right) - \frac{\partial g_{i4}}{\partial x^4},$$

where the Greek index ranges over the values 1, 2, 3. Hence show that the covariant components of the particle's acceleration in S are given by

$$\gamma_{\alpha\beta}\frac{d^2 x^\beta}{(dx^4)^2} = -\frac{\partial U}{\partial x^\alpha} - (c^2 + 2U)^{1/2}\frac{\partial \gamma_\alpha}{\partial x^4},$$

where $\gamma_{\alpha\beta}$ is defined in exercise 14 and

$$g_{44} = -(c^2 + 2U), \quad \gamma_\alpha = g_{\alpha 4}/\sqrt{(-g_{44})}.$$

(U, γ_α are the gravitational scalar and vector potentials respectively.)

Show that, in the case of the space-time metric appropriate to the rotating frame of exercise 15, the gravitational vector potential vanishes and the scalar potential is given by $U = \frac{1}{2}\omega^2 r^2$. Interpret this result in terms of the centrifugal force.

17. De Sitter's universe has metric

$$ds^2 = -A^{-1}dr^2 - r^2 d\theta^2 - r^2 \sin^2\theta \, d\phi^2 + Ac^2 dt^2,$$

where $A = 1 - r^2/R^2$, R being constant. Obtain the differential equations satisfied by the null-geodesics and show that along null-geodesics in the plane $\theta = \frac{1}{2}\pi$

$$a\frac{dr}{d\phi} = r(r^2 - a^2)^{1/2},$$

where a is a constant. Deduce that, if r, ϕ are taken to be polar coordinates in this plane, the paths of light rays in this universe are straight lines.

18. Einstein's universe has the metric

$$ds^2 = c^2 dt^2 - \frac{1}{1-\lambda r^2} dr^2 - r^2 d\theta^2 - r^2 \sin^2\theta\, d\phi^2,$$

where (r, θ, ϕ) are spherical polar coordinates. Obtain the equations governing the null-geodesics and show that, in the plane $\theta = \frac{1}{2}\pi$, these curves satisfy the equation

$$\left(\frac{dr}{d\phi}\right)^2 = r^2(1-\lambda r^2)(\mu r^2 - 1),$$

where μ is a constant. Putting $r^2 = 1/v$, integrate this equation and hence deduce that the paths of light rays in the plane $\theta = \frac{1}{2}\pi$ are the ellipses

$$\lambda x^2 + \mu y^2 = 1,$$

where (x, y) are rectangular cartesian coordinates. Show, also, that the time taken by a photon to make one complete circuit of an ellipse is $2\pi/(c\lambda^{1/2})$.

19. If the metric of space-time is

$$ds^2 = k\alpha\, dt^2 - \alpha^2(dx^2 + dy^2 + dz^2),$$

where α is a function of x alone and k is a constant, obtain the differential equations governing the world-lines of freely falling particles. If x,y,z are interpreted as rectangular cartesian coordinates by an observer and t is his time variable, show that there is an energy equation for the particles in the form

$$\tfrac{1}{2}v^2 - \frac{k}{2\alpha} = \text{constant}.$$

20. Explain why equation (23.6) remains valid in a gravitational field.

21. (r, θ, ϕ, t) are interpreted as spherical polar coordinates and time. A gravitational field is caused by a point electric charge at the

pole. Assuming that the space-time metric is given by equation (51.10) and that the 4-vector potential for the electromagnetic field of the charge is given by $\mathbf{\Omega} = (0,0,0,\chi)$, where $\chi = \chi(r)$, calculate the covariant components of the field tensor F_{ij} from equation (23.6) and deduce the contravariant components F^{ij}. Assuming that $J^i = 0$, prove that Maxwell's equations are all satisfied if

$$\frac{d\chi}{dr} = \frac{e}{r^2}.c^2\sqrt{(ab)},$$

where e is a constant.

Calculate the elements of the mixed energy-momentum tensor from the equation

$$T^i_j = \frac{1}{4\pi} F^{ik} F_{jk} - \frac{1}{16\pi} \delta^i_j F^{kl} F_{kl}$$

and write down Einstein's equations for the gravitational field. Show that these are satisfied provided

$$\frac{1}{a} = b = 1 - \frac{2m}{r} + \frac{\gamma e^2}{c^2 r^2},$$

where m is a constant.

Appendix

Bibliography

1. AHARONI, J., *The Special Theory of Relativity*, Oxford University Press.
2. BERGMANN, P. G., *Introduction to the Theory of Relativity*, Prentice-Hall.
3. EDDINGTON, A. S., *Mathematical Theory of Relativity*, Cambridge University Press.
4. EINSTEIN, A., *The Meaning of Relativity*, Princeton University Press.
5. FOCK, V., *Theory of Space, Time and Gravitation*, Pergamon.
6. LANDAU, L. D. and LIFSHITZ, E. M., *The Classical Theory of Fields*, Pergamon.
7. MCCONNELL, A. J., *Applications of the Absolute Differential Calculus*, Blackie.
8. MCCREA, W. H., *Relativity Physics*, Methuen.
9. MCVITTIE, G. C., *General Relativity and Cosmology*, Chapman and Hall.
10. MØLLER, C., *Theory of Relativity*, Oxford University Press.
11. PAULI, W., *Theory of Relativity*, Pergamon.
12. RAINICH, G. Y., *Mathematics of Relativity*, Wiley.
13. RINDLER, W., *Special Relativity*, Oliver and Boyd.
14. SCHRÖDINGER, E., *Space-Time Structure*, Cambridge University Press.
15. SOMMERFELD, A., *Electrodynamics*, Academic Press.
16. SPAIN, B., *Tensor-Calculus*, Oliver and Boyd.
17. SYNGE, J. L., *Relativity – The Special Theory*, and *Relativity – The General Theory*, North-Holland.
18. TOLMAN, R. C., *Relativity, Thermodynamics and Cosmology*, Oxford University Press.
19. WEBER, J., *General Relativity and Gravitational Waves*, Interscience.
20. WEATHERBURN, C. E., *Riemannian Geometry and Tensor Calculus*, Cambridge University Press.
21. WEYL, H., *Space, Time, Matter*, Dover.

Index

Aberration of light, 52
Acceleration, Lorentz transformation of, 51
Aether, 6
Affine connection, 99
Affinity, 99
  metric, 121
  symmetric, 111, 119
  transformation of, 99
Affinities, difference of, 101
Atomic explosion, 47

Bianchi identity, 114, 122
Biot–Savart law, 67

Cartesian tensor, 27
  covariance and contravariance of, 89
Charge, equation of continuity for, 59
  field of moving, 65
Charge density, proper, 60
Christoffel symbols, 120
Clock, coordinate, 168
  standard, 168
Clock paradox, 16
Compton effect, 55
Conjugate tensors, 93
Continuity, equation of, 59, 74
Coordinate lines, 82
Coordinates, curvilinear, 82
  geodesic, 114
  spherical polar, 81
Coordinates of an event, 144
Coordinate surfaces, 82
Copernicus, 5
Cosmical constant, 150
Cosmic ray particles, 42
  decay of, 16
Covariant derivative, 98
Covariant differentiations, commutativity of, 130
Curl, 35, 130

Current density, 59
  4-, 60, 172
  Lorentz transformation of, 77
Curvature scalar, 124
Curvature tensor, covariant, 121
  Riemann–Christoffel, 112
  symmetry of, 122
Curvature tensor for a weak field, 153

De Sitter's universe, 179
Dilation of time, 16, 38
Divergence, 31, 123
Doppler effect, 78

Einstein's equation, 46
Einstein's law of gravitation, 150
Einstein's tensor, 125, 150
Einstein's universe, 180
Electric charge, 60
Electric intensity, 62
  Lorentz transformation of, 65
Electromagnetic field tensor, 63, 171
Energy, equivalence of mass and, 46
  kinetic, 45
  particle's internal, 47
Energy current density, 73
Energy density in an electromagnetic field, 71
Energy-momentum tensor, 147
  electromagnetic field, 70
  kinetic, 76
Eötvös, 139
Equivalence, principle of, 139
Euclidean space, 10, 81, 84, 98, 117, 119, 121, 143, 150
Event, 8
  coordinates of, 144

Fitzgerald contraction, 14, 21, 142
Force, 2, 43
  centrifugal, 3, 21, 139, 142

183

Coriolis, 3, 21, 139, 142
  fictitious, 3
  4-, 44
  inertial, 138
  Lorentz transformation of, 48
  rate of doing work by, 45, 49
Force density, 68
Fundamental tensor, 28, 89, 115
  covariant derivative of, 103
Future, absolute, 19

Galactic masses, field due to, 140
Galilean law of inertia, 147
Galilean transformation, special, 13
General principle of relativity, 138
Geodesic, 126, 134, 147
  null, 128, 166
Gradient, 30, 87
Gravitational constant, 149, 154
Gravitational field of point charge, 181
Gravitational field outside a spherical mass, 159
Green's theorem, 72

Hamiltonian, 50, 79
Hamilton's equations, 50
Hypersphere, 136

Index, dummy, 26
  free, 26
  raising and lowering, 116
Inertial forces, 138
Inertial frame, 2
  local, 143
  quasi-, 151
Interval, 115
  proper time, 17, 144
  spacelike, 18
  timelike, 18
Interval between events, 145
Intrinsic derivative, 131
Invariant, 29, 87
  covariant derivative of, 102
  relative, 95
Invariant density, covariant derivative of, 105
  parallel displacement of, 104
Invariant field, 29, 87

Irreducible gravitational field, 143

Kronecker deltas, 25, 89

Lagrange's equations, 50
Laplacian, 124
Length, 15
Levi-Civita tensor density, 32, 94
  covariant derivative of, 106
Light cone, 20
Light pulse, wavefront of, 9
Light ray, gravitational deflection of, 166
Light waves, velocity of, 6
Lorentz force, 68
Lorentz transformation, general, 11
  inverse, 14
  special, 13

Mach's principle, 141
Magnetic intensity, 62
  Lorentz transformation of, 65
Mass, 2
  conservation of, 42
  density of proper, 74
  equation of continuity for proper, 74
  equivalence of energy and, 46
  inertial and gravitational, 141
  invariance of, 4
  proper, 42
  proper density of proper, 74
  rest, 42
  variable proper, 48
Maxwell's equations, 6, 61, 64, 172
Maxwell's stress tensor, 71
Mercator's projection, 136
Mercury, 164, 165
Metric, 84, 115
  form invariance of, 155
  Schwarzschild, 159, 174, 177
  spherically symmetric, 155
Metrical connection, 115
Metric of a conical surface, 134
Metric of a gravitational field, 147
Metric of a spherical surface, 85, 132, 136
Michelson–Morley experiment, 6
Minkowski, 9
Minkowski space-time, 10, 18

INDEX

Momentum, 3
 conservation of, 3, 41, 73, 77
 4-, 42
 Lorentz transformation of, 43

Newtonian potential, 149, 153, 159, 170
Newton's first law, 1, 8
Newton's law of gravitation, 153
Newton's laws, covariance of, 4
Newton's second law, 3
Newton's third law, 3, 44
Non-Euclidean space, 143

Operator, substitution, 26
Orbit, planetary, 55, 160
 equation of, 163
Orthogonal transformation, 23, 81

Parallel displacement of vectors, 97, 109
Particle, Hamiltonian for, 50
 internal energy of, 47
 Lagrangian for, 50
Particles, collision of, 3, 41, 54, 55, 56
Past, absolute, 20
Perihelion, advance of, 165
Photon, 53, 55
Physical space, 137
Planck's constant, 55
Poisson's equation, 154
Poynting's vector, 71
Present, conditional, 20
Privileged observers, 137
Product, inner, 31, 92
 outer, 91
 scalar, 31
 vector, 34, 130

Quotient theorem, 92, 111

Relative invariant, 95
Relative tensor, 94
 covariant derivative of, 108
Ricci tensor, 113
 divergence of, 124
Ricci tensor for a weak field, 153
Riemannian space, 84, 115
Ritz, 7

Rocket motion, 53

Scalar, 29, 87
Scalar potential, 61
Simultaneity, 15
Sirius, companion of, 171
Space-time continuum, 144
Special principle of relativity, 5
Spectral lines, gravitational displacement of, 168
Summation convention, 25
Synchronization of clocks, 7, 144

Tangent, unit, 126
 zero, 128
Tensor, Cartesian, 27
 contraction of, 31, 91
 contravariant, 88
 covariant, 88
 covariant derivative of, 103
 fundamental, 28, 89
 mixed, 88
 parallel displacement of, 97
 rank of, 27
 relative, 94
 skew-symmetric, 28, 90
 symmetric, 28, 90
Tensor density, 32, 94
 covariant derivative of, 106, 108
 Levi-Civita, 32, 94
Tensor equation, 29, 90
Tensor field, 30
Tensor product, covariant derivative of, 104
Tensors, addition of, 27, 90
 multiplication of, 27, 90
Tensor sum, covariant derivative of, 103
Time, absolute, 13

Vector, axial, 34, 35
 Cartesian, 27
 contravariant, 86
 covariant, 87
 covariant and contravariant components of, 116, 134
 displacement, 26
 free, 86
 infinitesimal displacement, 86
 magnitude of, 31, 118

Vector field, 87
Vector multiplication, laws of, 34
Vector potential, 61
   4-, 61
Vector product, 34, 130
Vectors, angle between, 31, 118
   orthogonal, 31, 118
   scalar product of, 31, 118
Velocities, composition of, 53

Velocity vector, 38
   4-, 39
   Lorentz transformation of, 40

Weak gravitational field, 151
Wilson cloud chamber, 54
World-line, 18
World-lines of free particles, 146